CHERT VARIETIES IN THE CABALLOS FORMATION

Top row, left to right: novaculite, blotchy novaculite, and brecciated slump-deformed chert. Middle row: subnovaculite, mottled red-green chert. Bottom row: banded black-green chert and green chert. Scale in inches.

The Caballos Novaculite, Marathon Region, Texas

Earle F. McBride
The University of Texas at Austin, Austin, Texas

and

Alan Thomson
Shell Development Company, Houston, Texas

THE GEOLOGICAL SOCIETY OF AMERICA

SPECIAL PAPER 122

Copyright 1970, The Geological Society of America, Inc.
Library of Congress Catalog Card Number: 74-98014
S.B.N. 8137-2122-9

Published by
THE GEOLOGICAL SOCIETY OF AMERICA, INC.
Colorado Building
P.O. Box 1719
Boulder, Colorado 80302

Printed in the United States of America

*The printing of this volume has been made possible
through the bequest of
Richard Alexander Fullerton Penrose, Jr.,
and by grants from the National Science Foundation,
the Shell Development Company, and the University of
Texas at Austin.*

Acknowledgments

It is a pleasure to acknowledge the courtesy of the following ranchers for granting access to their property: John Catto, David Combs, Guy Combs, William Donnell, Gage Holland, Tom Leary, John McNutt, Travis Roberts, and Steve Stumberg.

Keith Moreland of the Electron Microscope Laboratory of The University of Texas at Austin did the electron microscopy, and Daniel Schofield of the Mineral Studies Laboratory of The University supervised the chemical analyses.

Helpful discussions were given during the investigation by S. P. Ellison, Jr., J. F. Houser, and J. D. Powell. Parts of the manuscript were read and constructively criticized by V. E. Barnes, W. C. Bell, Bruce Boyer, D. H. Eargle, H. P. Eugster, R. L. Folk, P. B. King, L. S. Land, J. D. Powell, Gordon Rittenhouse, and Raymond Siever.

The National Science Foundation (Grant GP-4651) supported McBride's work, for which grateful acknowledgment is made.

Final preparation of the manuscript was made possible by time provided by a special faculty assignment to McBride, authorized by the Dean of the College of Arts and Sciences of The University of Texas at Austin. Part of the cost of publication of this report was borne by the NSF grant, by the Shell Development Company, Houston, Texas, and by a generous grant from the Ed Owen-George Coates Fund of the Geology Foundation, The University of Texas at Austin.

Contents

Abstract 1

PART 1. STRATIGRAPHY

Introduction 5
Previous work 10
Present work 13
    Lower chert member 14
    Lower novaculite member 16
    Lower chert and shale member 17
    Upper novaculite member 19
    Upper chert and shale member 19
Local Stratigraphy 20
    Marathon Basin 20
    Persimmon Gap and Rough Creek 24
    Old Jones Ranch 28
    Solitario Uplift 30
Problem of the lower contact of the Caballos Formation 31
Problem of the upper contact of the Caballos Formation 34
Fossils and Age 36
Discussion 39

PART 2. PETROGRAPHY AND ORIGIN

Introduction 45
Petrography 45
    General 45
    Novaculite 46
    Subnovaculite 56
    Tan chert 56
    Green chert 57
    Gray and black chert 58
    Brown chert 59
    Blue and red chert 59
    Banded and mottled chert 59
    Shale 60
    Sandstone 62

Conglomerate ... 63
Calcarenite ... 64
Comparsion with previous studies ... 64
Areal and vertical stratigraphic differences ... 65

Sedimentary and diagenetic features ... 66
   Pseudo ripple marks ... 66
   Turban concretions ... 68
   Pits and mounds ... 69
   Trace fossils ... 69
   Knobby bedding surfaces ... 70
   Liesegang bands ... 71
   Geopetal fabric ... 71
   Slumped and MnO_2-stained beds ... 71
   Iron-oxide nodules ... 73
   Miscellaneous ... 73

Petrogenesis ... 74
   Review of origins of other bedded chert formations ... 74
   Review of hypotheses of origin of the Arkansas and Caballos novaculites ... 75
   Summary of facts pertinent to the origin of the Caballos novaculite ... 79
   General interpretation of present article ... 80
   Comparison with previous interpretations ... 81
   Special problems ... 85
      Depth of water ... 85
      Purity of novaculite ... 87
      Red *versus* green shale ... 88
      Rate of deposition ... 89
      Spicule orientation ... 90
      Chemical diagenesis ... 90

Paleogeography and provenance ... 92
Geologic History ... 94
References cited ... 97
Plates ... 105
Index ... 125

FIGURES

1. Index map of study area ... 6
2. Stratigraphic section of Paleozoic rocks, Marathon region ... 7
3. Outcrop and locality map of the Caballos Formation ... 8
4. Stratigraphic subdivisions and terminology, Caballos Formation ... 12
5. Isopach maps of the Caballos Formation ... 15
6. Stratigraphic cross section of the Caballos Formation ... 21
7. Stratigraphic cross section of the Caballos Formation ... 23
8. Stratigraphic cross section of the Caballos Formation in Marathon Basin, Solitario Uplift, and Persimmon Gap ... 26
9. Stratigraphic cross section of the Caballos Formation, Marathon Basin ... 27
10. Stratigraphic section and terminology of Caballos at Rough Creek ... 29

11.	Diagrammatic cross section showing possible relations of time lines to novaculite and chert and shale members	41
12.	Cross section of pseudo ripple marks in novaculite	68
13.	Orientation of pseudo ripple marks in novaculite	68

PLATES

1. Chert varieties in the Caballos Formation
2. Outcrops of Caballos novaculite members
3. Outcrops of Caballos Novaculite
4. Chert with thin shale interbeds
5. Bedding details of chert and novaculite
6. Upper contact of the Caballos Formation and slump deformed chert bed
7. Texture of novaculite and subnovaculite in thin section
8. Texture of novaculite in thin section and etched surface
9. Electron micrographs of novaculite
10. Electron micrographs of novaculite
11. Texture of green, gray, and blue-black chert in thin section
12. Texture of shale, mottled brown chert, and sandstone in thin section
13. Structures in sandstone beds
14. Texture of radiolarian chert and shale, calcarenite, chert-pebble conglomerate, and novaculite
15. Radiolaria from the Caballos
16. Pseudo ripple marks in novaculite and knobs of diagenetic origin in chert
17. Iron-oxide concretion and turban concretions
18. Pits and mounds on novaculite bedding planes

TABLES

1.	Stratigraphic distribution of faunal elements	38
2.	Composition and texture of Caballos novaculite samples	48
3.	Chemical analyses of Caballos novaculite samples	53
4.	Description of chert textures in electron photomicrographs	54
5.	Mineralogy of shale samples determined by X-ray diffraction	62
6.	Origins proposed for the Arkansas novaculite	76
7.	Rate of accumulation of marine sediments	89

Abstract

The Caballos Novaculite is composed predominantly (90 percent) of bedded chert; the widespread distribution of two novaculite (milky white chert) lithosome marker units permits the formation to be divided into five members. From the base up, the members are listed with maximum thicknesses and rock type as follows: lower chert, 15 feet, light brown and off-white chert with shale partings; lower novaculite, 150 feet; lower chert and shale, 200 feet, chert of many different colors (chiefly green) and shale partings, and several beds of calcarenite; upper novaculite, 400 feet; and upper chert and shale, 400 feet, chert of many different colors and shale partings. Minor amounts of red and green shale, pebble conglomerate, sandstone and calcarenite are in the chert and shale units.

The Caballos Novaculite in the Marathon Basin is a lens-shaped unit from 100 to 700 feet thick, based on 26 measured sections. The Caballos is conformable with the underlying Maravillas Formation of Late Ordovician age and with the overlying Tesnus Formation of Mississippian to Early Pennsylvanian age; and, therefore, probably includes rocks of Silurian, Devonian, and possibly Mississippian age. The novaculite members may be time-stratigraphic units in the center of the Marathon Basin.

Chert formed by the diagenetic alteration of opaline skeletal particles, chiefly sponge spicules and Radiolaria. Quartz in the form of subsequent grains of microquartz ($<35~\mu$) and lesser megaquartz is the only silica phase present.

Color varieties of chert differ in fabric of microquartz and content of pigmenting impurities; green chert contains abundant illite; tan, brown, and black chert contain iron and manganese oxides and organic matter; red chert contains hematite; blue chert contains apatite. Mottled chert beds formed by submarine slumping.

Novaculite (snow-white chert) owes its lack of color to the absence of detrital impurities and chemical pigments, and its milkiness to the dispersion of light by minute water inclusions (Folk, 1965) and reflecting faces of microgranular quartz crystals. Novaculite has a distinct microscopic fabric; subspherical specks up to 200 μ in diameter composed of microquartz grains up to 10 μ long are scattered in a matrix of slightly larger uneven-grained microquartz grains, 5 to 25 μ long; the specks are probably relict Radiolaria. Skeletal ghosts of spicules visible in ordinary light range from 0 to 80 percent of novaculite beds, but average 10 percent. Radiolaria are more abundant than spicules in colored chert beds.

The Caballos was deposited in a deep-marine trough adjacent to peneplaned land masses; rate of accumulation (assuming 50 percent compaction) was from 0.1 to 0.5 mm/1000 years. Terrigenous clay was absent during accumulation of proto-novaculite, but was supplied occasionally during accumulation of colored chert and shale beds. Green and red (originally brown?) colors of clay shale reflect original differences in bottom conditions during deposition. Quartz silt in chert includes wind-blown and storm-deposited grains; sandstone, calcarenite, and conglomerate beds were deposited by rare turbidity currents.

Depositional fabrics of sediments were modified by lithification processes, burrowing infauna, escaping gases, and submarine slumping. Chertification occurred largely by solution of opal and precipitation of crystalline silica (cristobalite or low-quartz) from pore solutions relatively soon after deposition.

Part 1: Stratigraphy

EARLE F. MCBRIDE

Stratigraphy

INTRODUCTION

The Caballos Novaculite[1] is one of the most distinctive formations exposed in the Marathon and Solitario regions of west Texas (Fig. 1). The Caballos has attracted considerable attention because of the unusual character of novaculite (white chert) and problems relating to its origin, the abundance of chert of other colors in the formation (Pl. 1), and because it is the chief ridge-former and best delineates major structural features in the Marathon and Solitario regions. Part 1 of this report describes the stratigraphy of the Caballos as revealed in outcrop. Part 2 describes the petrography and origin as interpreted from outcrop and laboratory studies.

The Caballos Novaculite is part of a thick sequence of Paleozoic sedimentary rock (Fig. 2) that was deposited in the Ouachita geosyncline in trans-Pecos Texas. The Marathon geosyncline, as this part is commonly called, received sediment throughout most, if not all, of the interval from Late Cambrian through Early Pennsylvanian time (Thomson and McBride, 1964). The Paleozoic fill of the geosyncline was folded and thrust-faulted as the result of crustal shortening that culminated during Late Pennsylvanian time (King, 1937).

The regional setting of the deformed Paleozoic rocks of the Marathon Basin and adjacent areas is described by King (1937), Flawn (1964), and Daly (1964). The similarity of the Paleozoic

---

[1] For Measured Sections of Caballos Novaculite, Marathon Basin, Texas, order NAPS Document 00311 from ASIS National Auxiliary Publications Service, c/o CCM Information Sciences, Inc., 22 West 34th Street, New York, New York 10001; remitting $1.00 for microfiche or $3.00 for photocopies. Checks may be made payable to ASIS-NAPS.

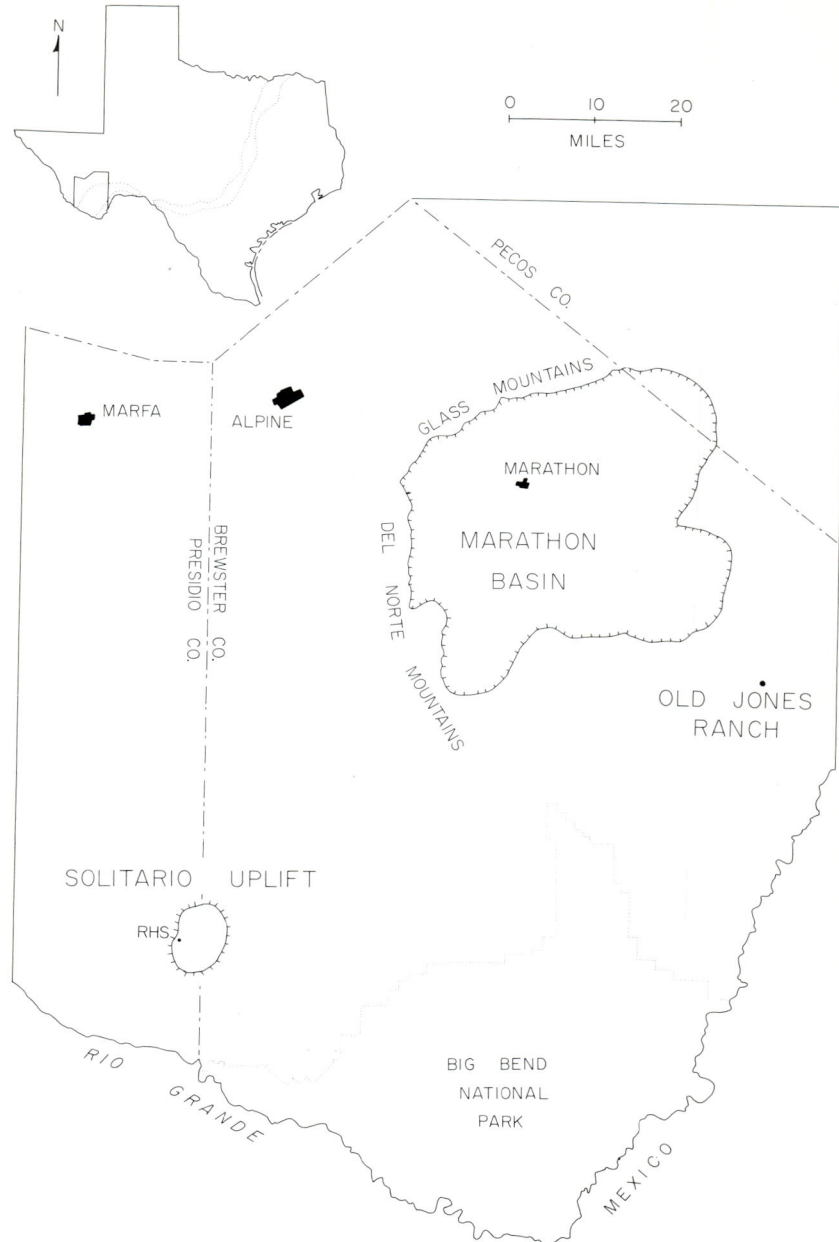

Figure 1. Index map of study area. Dotted belt across Texas (upper left) shows position of the deformed rocks of the frontal Ouachita system in subsurface (*after* Flawn and others, 1961). Small outcrops of Caballos occur at the old Jones Ranch and Persimmon Gap.

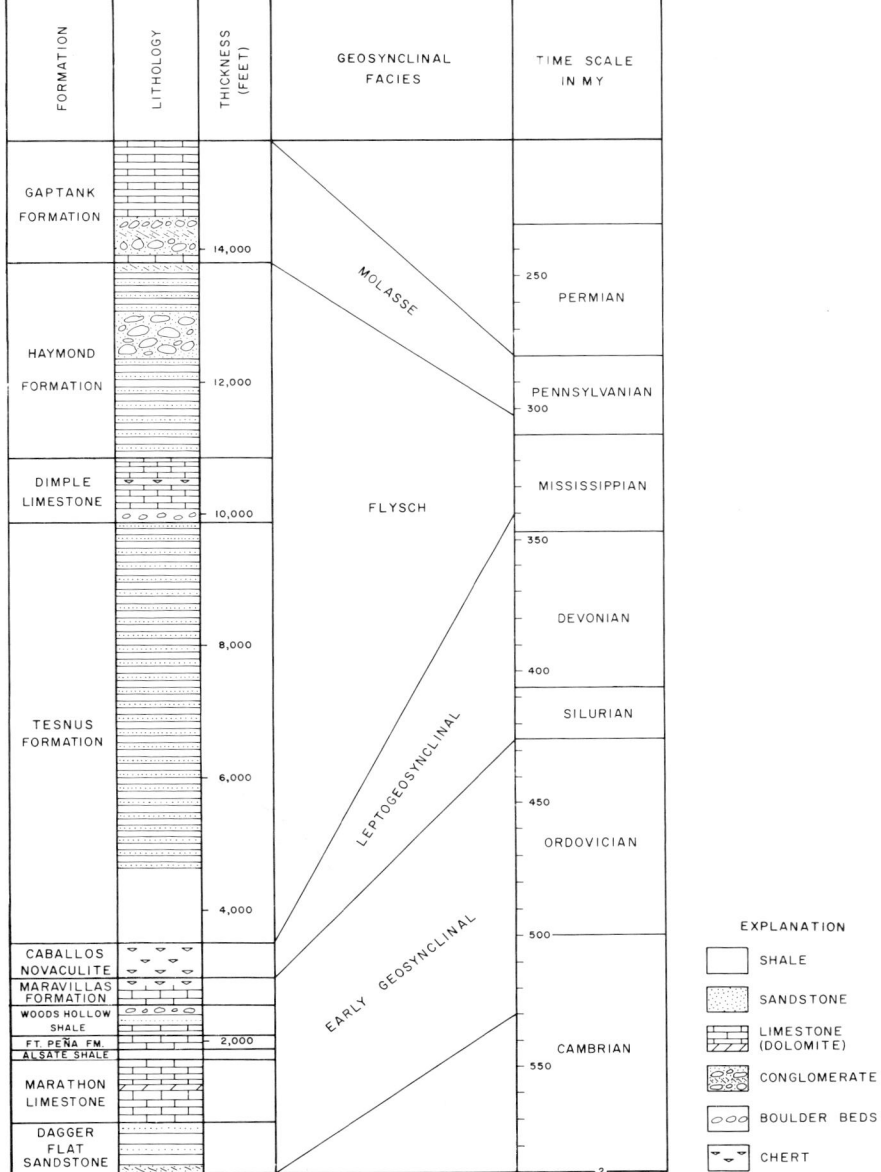

Figure 2. Stratigraphic section of Paleozoic rocks exposed in the Marathon region.

sequence in the Marathon region with that in the Ouachita Mountains in Arkansas and Oklahoma has been noted for many years and supports the contention that they belong to the same genetic-structural unit (Flawn, 1964). The outcrops in which the Caballos was studied, as part of this investigation, are at the southwestern end of the known limits of the Ouachita structural belt.

The Caballos Novaculite crops out extensively in the Marathon Basin (Fig. 3) and in several outlying areas to the south and east

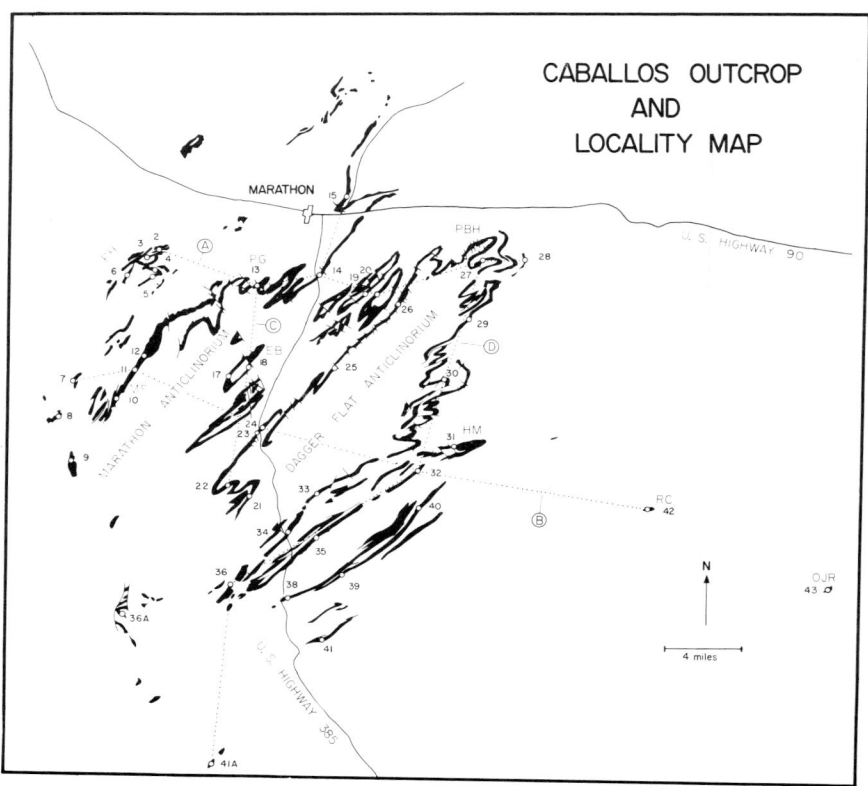

Figure 3. Outcrop and locality map of the Caballos Formation in the Marathon Basin. Numerals refer to measured sections and other data localities. Abbreviated letters are as follows: PH = Payne Hills, PG = picnic grounds at old Fort Peña, EB = East Bourland Mountain, MS = Monument Springs, PBH = Peña Blanca Hills, HM = Horse Mountain, RC = Rough Creek locality, OJR = old Jones Ranch locality. Dotted lines show positions of cross sections A (Fig. 6), B (Fig. 7), C (Fig. 8), and D (Fig. 9). Outcrop after King (1937).

(Fig. 1). The Marathon Basin is a topographic low situated on a structurally high part of the Ouachita structural belt; it has dimensions of 50 miles in a northeast-southwest direction and 35 miles in a northwest-southeast direction. Folds in deformed Paleozoic rocks trend northeast-southwest in the basin, and are exposed through a veneer of Quaternary alluvium. The basin is rimmed by Permian rocks to the northwest (Glass Mountains) and by Cretaceous rocks in other directions.

The largest outlying outcrop is in the Solitario Uplift, a circular area domed by igneous intrusion (Herrin, 1958), 40 miles southwest of the Marathon Basin. Small exposures are also present between Persimmon Gap and Dog Canyon in the Santiago Mountains, at Rough Creek, and at the old Jones Ranch.

The Caballos is a sequence of conspicuous novaculite (snow-white chert) and green, gray, brown, tan, and off-white chert beds (Pl. 1) with lesser green and red clay shale and siliceous shale. A few beds of limestone (calcarenite), sandstone, and silicified granule conglomerate are also present. The formation has a maximum thickness of 700 feet, but is generally between 200 and 500 feet thick. It overlies the Maravillas Chert, the uppermost unit of an Ordovician succession of limestone, shale, chert, and conglomerate that has a total thickness of 2000 feet. The Maravillas contrasts strongly with the overlying Caballos because the bedded chert in the Maravillas is predominantly black. The Caballos is succeeded by the Tesnus Formation, the lowermost of three Carboniferous flysch units that have a total thickness of 12,000 feet.

Reconnaissance work on this study and a concurrent study of the Maravillas Chert began in 1960, but most of the field work was done in the summers of 1965 and 1966. The stratigraphy of the Caballos, presented herein, is based on 26 measured sections in the Marathon Basin plus measured sections at the old Jones Ranch, Persimmon Gap, and two in the Solitario Uplift.

The term novaculite has been used with different meanings by numerous authors. According to Griswold (1892, p. 2, 84), the Englishman, Richard Kirwan, in 1784 coined the anglicized word *novaculite* from the Latin *novacula,* a term used by Cicero, Petronius, Celius, and others, for first a sharp knife and then razor, then by Linnaeus to denote a fine quality of whetstone. Colloquial use of novaculite extended its meaning from that of a whetstone to the rock from which whetstones were made. The term novaculite was applied first in the Arkansas region by Schoolcraft (1819, p. 183), who referred to the mineral novaculite at Hot Spring, Arkansas, then by Featherstonhaugh (1835, p. 69), who applied the term to whetstone raw material; Griswold (1892, p. 2) urged that novaculite be used in the petrographic sense for a variety of chert. Color was not implied in the original petrographic use by Griswold and others. However, when "Arkansas novaculite" became established as a formation name, novaculite was associated with the predominant chert type of the formation, the snow-white chert. Use of novaculite as a petrographic term was extended to the Marathon region by Baker (*in* Udden, Baker, and Böse, 1916), who commonly applied novaculite to white chert only. This usage was extended by King (1930, 1937). Novaculite is applied in this report to milk-white chert; it is nonporous (dense of most descriptions), megascopically homogeneous, and generally has duller luster than porcelain.

The colors of chert in the Caballos Formation (Pl. 1) represent a wide spectrum; white, off-white, green, gray, brown, black, red, and blue colors are present. Many beds are banded by several colors and grade into lighter or darker shades along strike. Chert beds that are off-white, vitreous, or color-mottled are called chert rather than novaculite in this report. Subnovaculite is applied to beds that are intermediate in character between chert and novaculite. Because the colors are gradational, this usage is necessarily subjective.

PREVIOUS WORK

The earliest mention of rocks of the Caballos Formation is in a report on a reconnaissance study by von Streeruwitz (1891, p. 686), in which he described ridges composed of "quartz and quartzite, strongly metamorphosed limestone, and semifused siliceous conglomerations." Additional observations were made by Hill (1900) and Udden (1907). The first outline of the stratigraphy of the Marathon region was based largely on the work of Baker (Udden, Baker, and Böse, 1916; Baker and Bowman, 1917). The Caballos was named in 1916 and amended in 1917 by Baker. The most detailed picture of the stratigraphy of the Caballos in the Marathon Basin was presented by P. B. King (1937) in his classic Professional Paper on *The Geology of the Marathon Region, Texas*. Earlier comments on the Caballos by King appeared in 1930 and

1931. Revisions of King's stratigraphy were proposed by Berry and Nielsen (1958), Bennett (1959), and Fan 1964).

Notes on Caballos outcrops on the old Jones Ranch are given by Berry and Nielsen (1958), and on the Caballos in the Solitario Uplift by Powers (1921), Baker (*in* Van Waterschoot van der Gracht, 1931), Herrin (1958), and Berry and Nielsen (1958). Maps of the Solitario Uplift showing the distribution of the Caballos were made by Sellards and others (1930), Wilson (1954), and Herrin (1958).

Figure 4 summarizes the various stratigraphic subdivisions of the Caballos and adjacent strata that have been made by previous investigators and the subdivisions used in this report. The marked facies changes displayed by bedded chert units is responsible for much of the confusion that has arisen.

The early history of the Caballos nomenclature is well summarized by Berry and Nielsen (1958, p. 2254-2255):

> Overlying the Ordovician succession in the Marathon region is a sequence of white novaculites, varicolored cherts, and shales. This sequence was originally divided by Udden, Baker and Bose (1916, p. 39, 41) into two formations, the Caballos novaculite below, and the Santiago chert above. The Caballos novaculite of those authors was described as follows: "the lower 40 feet is of rather thin-bedded light brown chert with some layers of snow white novaculite. The upper 50 feet is massive-bedded, ripple-marked and much fractured white novaculite." The Santiago chert was described as "thin-bedded, banded, or ribboned, of dull shades of practically every color but mostly green," and thought to unconformably overlie the Caballos novaculite. In 1917 (p. 93) Baker and Bowman rejected the name "Santiago chert" because, "later work by the senior author appears to indicate that they (Caballos and Santiago) are really one formation." The rocks formerly named "Santiago chert" were included with those originally named "Caballos novaculite" to make one unit, the Caballos novaculite, with a new difinition. The type locality of the new Caballos was designated as Horse Mountain. . . . P. B. King (1937, p. 48) stated that his work "afforded complete confirmation of this latter interpretation." King made a detailed study of the formation and divided it into five members—a lower chert member at the base, a lower novaculite member, a middle chert member, an upper novaculite member, and an upper chert member at the top.

From their regional interpretation of the Caballos, Berry and Nielsen (1958, p. 2258) proposed to revise the original terminology of Baker (*in* Udden, Baker, and Böse, 1916). They report that

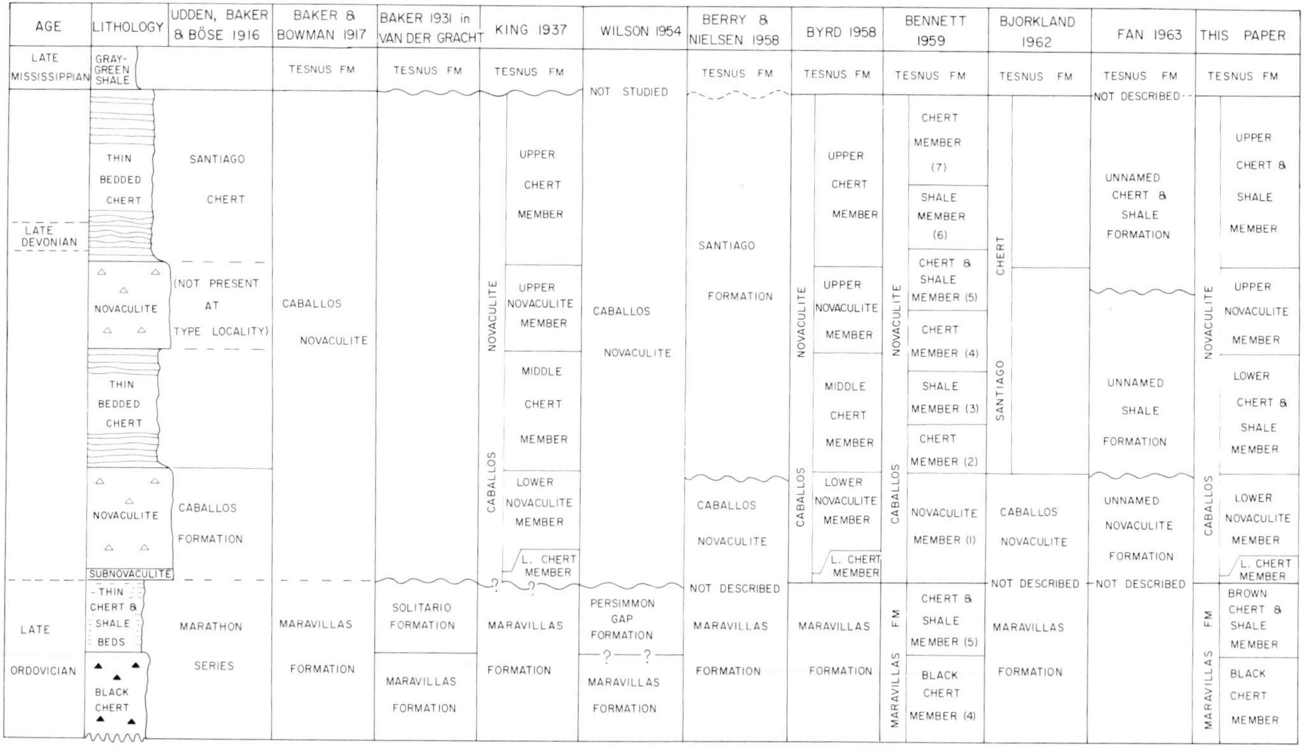

Figure 4. Stratigraphic subdivisions and terminology of the Caballos and adjacent strata.

the lower two members of the Caballos Novaculite of Baker and Bowman (1917) and King (1937) pinch out to the south and are uncomformably overlain by the upper three members. They apply Caballos Novaculite to the first two members of King (1937) and Santiago Formation to the three upper members of King. The pinchout of the lower two members was first described by King, but not attributed to erosion by him.

Bennett (1959) mapped an area that includes parts of East Bourland and Simpson Springs Mountains, and established six members in the Caballos. He established these informal members for his map area only.

Fan (1964), in an abstract, proposed to subdivide the Caballos of King into three new unnamed formations: a lower novaculite formation, which is uncomformably overlain by a shale formation, which is succeeded comformably by a chert and shale formation. Fan believed the units are genetically unrelated.

Until recently, most workers have reported the Caballos to be bounded above and below by unconformities. The nature of the formation contacts will be treated in detail following a presentation of first-order stratigraphic details.

PRESENT WORK

The stratigraphic framework of the Caballos in the Marathon Basin was superbly laid out by King (1937) and documented with 14 measured sections. The 26 measured sections made in the basin during this study provide a more detailed picture commensurate with the goals of the project. My work confirms the soundness of King's work in most details, and his report is a necessary preface to this one. The five members established by King are employed as the most useful stratigraphic subdivision. However, I depart from King's interpretation that the Caballos is bounded by unconformities, and reject the interpretation of Fan (1964) and Berry and Nielsen (1958) that unconformities exist within the formation.

Throughout most of the study area the Caballos can be divided conveniently into members using the two novaculite lithosomes as marker units. On this basis, King established five informal members, but did not describe their contacts in detail. The character of the five members, modified slightly, is described in more detail herein. In ascending order they include: (1) lower chert, (2) lower novaculite, (3) lower chert and shale, (4) upper novaculite, and (5) upper chert and shale. Where the novaculite members are absent, the chert and shale members cannot be separated.

A comment should be made about the deformed rocks of the region insofar as it bears on a stratigraphic analysis. The Caballos is the most competent stratigraphic unit of the Paleozoic fill of the Ouachita geosyncline in trans-Pecos Texas. Many outcrops of the formation are thrust-fault slices that have moved several miles from their original position. Exposures of the novaculite members are excellent in spite of locally severe jointing, but the thin-bedded chert and shale members generally form poor outcrops and locally are intensely drag-folded. Thickness measurements (made with a steel tape and 5- and 6-foot jacob staffs) of the well-exposed novaculite members were made with reproducible results, but those of the other members may frequently be in error 10 percent or more, because of failure to correct for drag folds or faults that are concealed by colluvium.

The five informal members of the Caballos are described below in ascending order.

Lower Chert Member

This unit is composed of white, tan, and gray-white chert and subnovaculite beds that are generally from 0.5 to 6 inches and locally, 12 inches thick. The member differs considerably in character across the area of study. In places, siliceous shale beds (ranging from paper-thin partings to 0.5 inch thick) separate chert beds, and the bedding planes of the chert weather to smooth surfaces that locally are dimpled by small pits or knobs. Where chert rests on chert with no intervening shale, the beds are tightly welded. The bedding planes between chert beds commonly have been altered to conspicuous stylolite surfaces, many of which have stylolite sutures with amplitudes up to 0.5 inch. Shale beds are weathered, and commonly their former presence is shown only by a gap between chert beds.

The member has a maximum thickness of 15 feet in the northern part of the basin (Fig. 5). It cannot be distinguished along the western edge of the basin, where the lower novaculite (member 2) is absent, and is not present in areas outside the basin. Chert beds in the member differ in places only slightly from typical novaculite, although the top of the member is generally easily differentiated by its off-white color or blotchy texture from more homogeneous novaculite. The member overlies what I refer

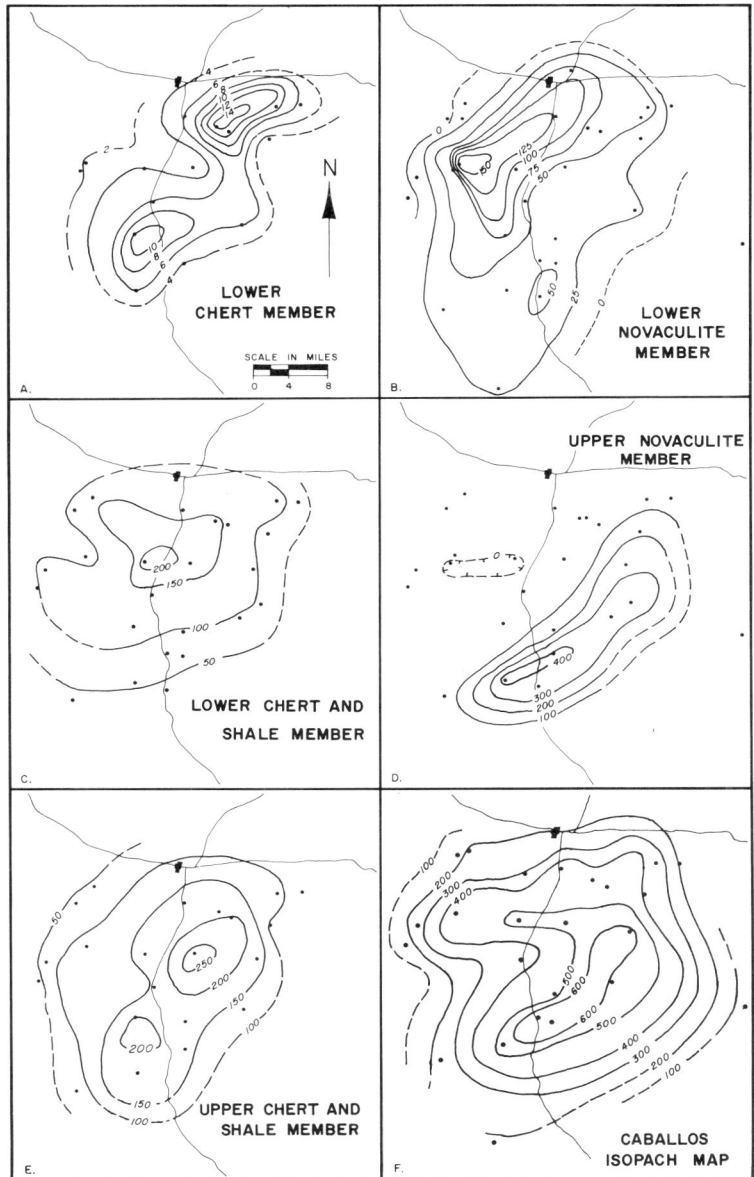

Figure 5. Isopach maps of the Caballos Formation and its members in the Marathon Basin. Contour values are in feet; contour intervals are not equal on all maps. Dots show data localities. Maps uncorrected palinspastically. Refer to Figure 3 for geography of Marathon Basin.

to as the brown chert and shale member of the Maravillas Formation. The lithology immediately below the bed designated as basal Caballos is a thin layer or parting of shale which generally caps a brown siliceous shale or shaly chert bed. At several localities a black, chertified-limestone and chert-pebble conglomerate bed from 2 to 14 inches thick is cemented to the basal chert bed of the Caballos; the conglomerate is assigned to the Maravillas Formation. Generally, the upper beds of the Maravillas are covered by colluvium.

In many areas this member is difficult to distinguish from the overlying novaculite member, particularly where shale partings are scarce within the chert beds. The possibility exists that non-novaculite chert beds may have been produced by secondary alteration of novaculite, where ground water had access to novaculite along shale partings. Thin-section study reveals a similarity between the lower chert and novaculite in places and reinforces this probability. Nevertheless, the member was designated where it could be recognized.

The relation of my lower chert member to King's lower chert member is uncertain. King (1937, p. 49) described his chert member in the Marathon anticlinorium as ". . . brown, vitreous, splintery, thick-bedded chert." From various comments in his report, it seems that his chert member may include what most subsequent workers have considered uppermost Maravillas. This problem is discussed in more detail later.

Lower Novaculite Member

This unit (Pls. 2A, 2B, 3B) is composed of typical novaculite and is the most homogeneous member of the formation. A few beds, pods, or lenses of subnovaculite and off-white and tan chert are locally present. At old Fort Peña, joint surfaces are coated with sufficient hematite to impart a pale red to orange color to the member when viewed from a distance. Local concentrations of MnO_2 along joints and bedding planes is rarer, but equally conspicuous. The novaculite has been altered to a blotchy-colored and blotchy-textured white to gray-white chert or subnovaculite in areas of greatest deformation.

The novaculite forms a compact, tightly welded unit that commonly forms steep cliffs (Pl. 2A). Bedding planes are generally spaced from 2 to 16 inches apart, and many have been modified into dentate surfaces of stylolites (Pl. 5B). Paper-thin shale partings occur in the upper part of the member. Where shale is weathered away, smooth bedding surfaces on chert are exposed which are generally featureless or which have broad warps with a few millimeters of relief. Unusual and rare bedding-surface features are knobby protuberances of probable diagenetic origin, bumps that are burrow entrances of benthonic animals, and pits or mounds probably formed by escaping gas. More abundant, but still uncommon, are rugose surfaces that resemble ripple marks. Problems concerning the origin of these bedding structures are discussed in Part 2.

This member has a maximum thickness of 122 feet on the western limb of the Marathon anticlinorium, but is absent at the extreme western margin of the basin (Fig. 5).

The lower novaculite member is overlain by interbedded chert and shale of member 3. The contact is concealed by a talus of novaculite rubble. In several places, novaculite occurs a few inches below beds of gray, tan, or green chert, or in places below a chertified sandstone of the overlying member, showing that the boundary is abrupt. The member has the same upper boundary as King's lower novaculite member, but the lower boundary may not be comparable for reasons mentioned later.

Lower Chert and Shale Member

This member is composed of thin green chert beds intercalated with green and gray siliceous shale (Pls. 4B, 5A) and with lesser fissile green and red clay shale. Chert beds with colors other than green (tan, dark brown, black, white, speckled gray-white, smoky gray, vitreous blue, and red chert) are locally present, as are chertified calcarenite, sandstone, and granule conglomerate. Proportions of the rock types differ with lateral facies changes, but dull green or banded light-green and gray chert comprise about half the unit.

Chert beds range from a millimeter to 16 inches thick, but most are between 1 and 3 inches. They are separated by shale or siliceous shale layers from partings to beds 0.5 inch thick. Bedding planes are generally smooth undulose surfaces; a few have trace fossil marks of pits, bumps, or trails of low relief; locally, knobby surfaces with relief of 0.5 inch are present. Trails of animals are rarer than the circular cross sections of vertical burrows. In a sequence that is essentially all chert beds separated by shale partings, the undulose bedding planes appear continuous from a distance. Closer inspection shows that some partings die out laterally and that two chert beds merge into one. The undulosity (Pl. 5B) of bedding planes is sufficient to change the thickness of beds to a third of their maximum thickness over a distance of a few feet. The thick part of one bed is succeeded by the thin part of another to even up major irregularities. Another peculiarity is typical of thick gray chert beds; they have undulose tops and bottoms and poorly defined internal laminae. Where laminae are visible, they are undulose and resemble ripple marks in style, but not in regularity. Layers and lenses of coarse siltstone and fine sandstone from 1 to 5 mm thick are common, but inconspicuous. About half these layers are graded.

Beds of sandstone, calcarenite, and chert granule conglomerate, the latter two beds partly to completely silicified, in layers from 1 to 8 inches thick are unusual and sparse rock types. Several sandstone beds of this type are widespread in the basal few feet of the member. Most sandstone and half the conglomerate beds are graded; some have unequal thickness along strike, and a few were found to pinch out within an outcrop.

Graded calcarenite beds at Localities 5 and 7 are in the lower chert and shale member. At Locality 5 are three calcarenite beds, 14, 15, and 18 inches thick (where measured), respectively. Lower in the member at Locality 7 is a bed 9 feet thick with granule-size chert clasts; the bed is composed of several sedimentation units.

Fissile red and green clay shale commonly form beds several feet thick and are most common at the top of the member. Conspicuous concentrations of MnO_2 occur along joint surfaces and pervade some chert and shale beds at many positions in the member.

The various chert and shale types are not uniformly interbedded through the section, but occur in distinct packages. For example, 10- to 20-foot intervals of thin-bedded green chert and sparse siliceous shale are succeeded by alternations of thicker-bedded gray or white chert and shale, then by units in which shale is equally abundant as chert. The lateral continuity of such units cannot be determined because exposures are not adequate. However, red and green shale beds near the top of the member have been found at widely separated outcrops in the central part of the basin (Figs. 6 and 8).

This member ranges from 40 to 200 feet thick (Fig. 5), and is identical to King's middle chert member. Because it is the lower of two similar members, and because of the abundance of shale and siliceous shale (estimated to be 10 to 30 percent of the member), it is herein named the lower chert and shale member. The shale-rich top of the member is overlain by the basal novaculite bed of the upper novaculite member. The lower bedding surface of this unit is commonly exposed, indicating it rested on a shale layer. The bedding surface in places has imprints of animal trails and burrow entrances.

The member is commonly covered by novaculite talus or a grassy slope (Pl. 2B), so that it is generally not well exposed.

Upper Novaculite Member

This unit is composed of novaculite similar to the lower novaculite member (Pls. 2B, 3A), but has more beds and lenses of subnovaculite and, locally, off-white chert than the lower unit. Beds of dark-colored chert are locally present, and welded bedding surfaces and stylolitic surfaces are also common. Shale partings are common in the upper few tens of feet of the unit, where most of the non-novaculite chert beds occur. Bedding planes are also more uneven in this part of the unit.

The upper novaculite is identical to King's upper novaculite member. It is thickest along the southeastern core of the Dagger Flat anticlinorium; at Locality 36 it has a maximum thickness of 430 feet (Fig. 5). It thins abruptly to the west, and throughout most of the Marathon anticlinorium is composed chiefly of a few beds of subnovaculite or off-white, speckled chert. This interval of subnovaculite also is designated upper novaculite member because it can be distinguished from adjacent strata.

The upper contact of the member generally is covered by novaculite talus, but a few exposures show that it is gradational over a few feet where novaculite, shale, and green chert are interbedded.

Upper Chert and Shale Member

This member is similar in lithology to the lower chert and shale member and is equivalent to King's upper chert member. It is composed chiefly of beds of dark-green, greenish-gray, greenish-brown chert in layers from 1 to 6 inches thick that are separated by shale layers from partings up to 1.5 inches thick. Chert beds of other colors are more abundant than in the middle member. An interval from 10 to 30 feet thick of red fissile clay shale with a few thin chert beds is a prominent unit from 10 to 60 feet from the top of the formation along the western limb of the

Marathon anticlinorium and eastern limb of the Dagger Flat anticlinorium. Away from this region, the red shale becomes siliceous and nonfissile and grades into nonred siliceous shale.

Chertified calcarenite beds are rare, but two to four granule conglomerate beds from 1 to 8 inches thick are present in many areas. The conglomerate beds are more abundant and thicker in the basal part of the Tesnus Formation, into which this member grades. Granule conglomerate beds are of uneven thickness along strike and commonly pinch and swell irregularly.

The upper chert and shale member is characterized by two additional features that commonly are related: mottled green-brown-black chert beds of uneven thickness that have lumpy bedding surfaces, and beds stained black by MnO_2. Bedding planes of the mottled beds have knobby, rounded, cauliflower-like lumps or dimpled depressions (Pl. 6B). Beds are generally dark green or gray-black, but have irregular blotches of red, green, or gray chert. Locally, beds have a brecciated appearance and are riddled by veins of white chalcedony that contrast markedly with darker colors of the host rock (Pl. 1). Most beds have granule- and pebble-size chert and limestone clasts (locally up to 2 feet long) as scattered grains in conglomerate layers or in conglomerate pods. At Locality 14, a bed has fragments of petrified wood (*Callixylon* sp.) up to 2 inches long. As described later, these layers are interpreted as beds deformed by soft-sediment slumping. At least one such bed from 6 to 30 inches thick has been found at most localities, even outside the Marathon Basin, 30 to 40 feet from the top of the member, but as many as four beds occur at places. Mottled beds have been found from 15 to 120 feet above the base of the member.

Some mottled beds are stained locally by MnO_2, either along bedding and joint surfaces, or entirely impregnating the bed to produce a coal-black rock. Small amounts of MnO_2 stain are common, but several areas are rich enough that manganese prospects have been worked.

This member is thickest along the Dagger Flat anticlinorium, where it has a maximum thickness of 260 feet (Fig. 5).

LOCAL STRATIGRAPHY

Marathon Basin

The 24 complete measured sections from the Marathon Basin provide the data for the stratigraphic framework presented here, and supplement the 14 sections presented by King (1937, Pl. 2). Eight of my sections were measured at or near the same places as King's. My thickness values differ slightly from King's, probably because of different interpretations of minor structural details.

Facies changes are illustrated by isopach maps of each member and of the formation (Fig. 5), and by four cross sections (Figs. 6-9). The maps and cross sections demonstrate the prominent lateral changes in thickness of members first documented by King (1937) and re-emphasized by Berry and Nielsen (1958). The maps are based on the present geometrical relations of outcrops. Measured sections at Localities 2, 7, 13, 14, 18, 27, 32, 36, and 41A are available. (See footnote, p. 5.)

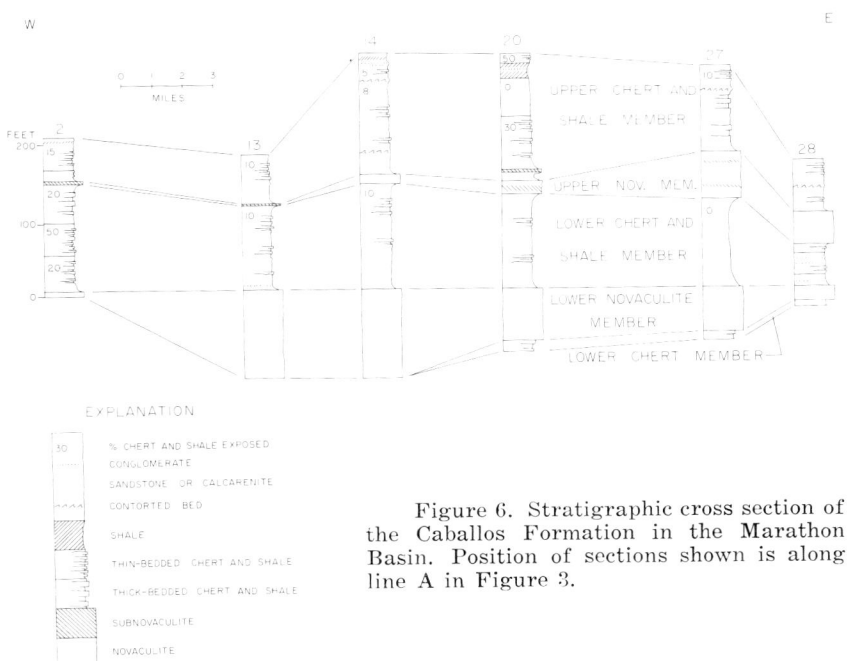

Figure 6. Stratigraphic cross section of the Caballos Formation in the Marathon Basin. Position of sections shown is along line A in Figure 3.

An accurate palinspastic reconstruction cannot be made because the amount of displacement that took place along thrust faults is unknown. If King's (1937, p. 133) calculated minimum values of crustal shortening are used to construct a palinspastic map, the isopach lines define a basin elongate in a northwest-southeast direction instead of in its present northeast-southwest direction.

The lower chert member can be recognized throughout most of the basin, but is absent at the margins. It is absent at some localities in the middle of the basin, although whether by facies change or by thrust faulting is uncertain.

The two novaculite members have prominent axes of greatest thickness that trend northeast-southwest. The two chert and shale members have axes of greatest thickness with a similar trend, but the axes are poorly defined. The axes of greatest thickness for members 1 to 4 show a progressive shift from northwest to southeast. This shows that the topography of the basin changed with time, insomuch as the axes of greatest thickness probably correspond to the positions of greatest basin depth during sedimentation. The lower novaculite is thickest in the Marathon anticlinorium, whereas the upper novaculite is thickest in the eastern flank of the Dagger Flat anticlinorium.

In several areas the formation differs in appearance from the majority of its exposures. The westernmost exposures are most strikingly different, because the novaculite members are only a few feet thick or are represented by a few beds of subnovaculite interbedded with shale. The following notes illustrate its character at the western edge of the basin.

Locality 2. The formation (217 feet thick) is chiefly interbedded light-brown to gray-white chert and siliceous shale. The lower novaculite member is 7 feet thick; the upper is represented by 4 feet of brecciated subnovaculite with shale partings (Fig. 6).

Locality 3. Interbedded chert and shale like Locality 2, but only 130 feet thick. Neither novaculite member can be identified.

Locality 5. The formation is 258 feet thick. The lower novaculite member is 18 feet thick, but the upper member is absent. A few beds of subnovaculite separated by green chert occur through 30 feet of section approximately where the upper novaculite member should be present.

Locality 6. In 180 feet of section there are a few widely separated beds of subnovaculite (Pl. 4A). Green chert and shale are poorly exposed.

Locality 7. The lower novaculite member is 18 feet thick, the upper is 39 feet thick, and the formation is 276 feet thick (Fig. 7).

Locality 8. The lower novaculite member is absent, and the upper is probably represented by 4 feet of severely brecciated and recrystallized white chert mottled by pods and layers of green, red, and brown. The formation is 131 feet thick. The lower green chert beds have spectacularly knobby bedding surfaces.

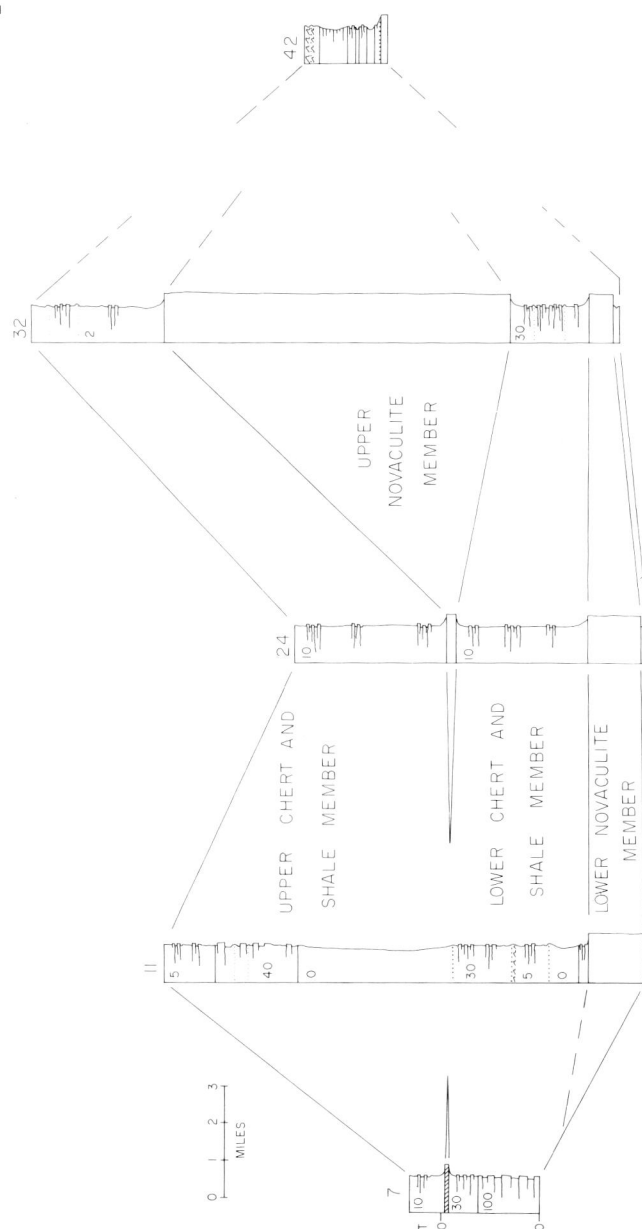

Figure 7. Stratigraphic cross section of the Caballos Formation in the Marathon Basin. See Figure 6 for explanation of symbols. Position of sections shown is along line B in Figure 3.

Locality 9. Approximately 200 feet of Caballos that is severely brecciated and faulted makes up the section. The basal 2 feet may be altered novaculite; it is blotchy-tan to orange-brown chert with dark-gray to blue-black pods and blobs. A chert pebble-cobble conglomerate bed, 4 feet thick, occurs in the upper part of the section; the largest clast is 18 inches long. Clasts are white, black, and green chert in a chert matrix that has similar colors.

Another area where the formation is atypical is along the western edge of the Dagger Flat anticlinorium, along the ridge south of Horse Mountain. The upper novaculite on Horse Mountain (Loc. 31) is composed of beds from 2 to 14 inches thick, between which shale partings are common. The member lacks the typical massiveness of most novaculite outcrops because distinct novaculite beds stand out in relief where the shale is removed during weathering (Pl. 3A). To the southwest, shale beds increase in thickness and together with thin, green, light-brown, and black chert beds make up bench-forming units that are locally mappable. The member in places is composed about equally of novaculite, subnovaculite, and tan or light-grey chert. Two miles northeast of U.S. Highway 385 along this ridge are two poorly resistant chert and shale units that separate the upper novaculite into three units. A mile south of this ridge, Bjorklund (1962) mapped a small area and established two upper novaculite members separated by a black chert member. At Maravillas Gap (Loc. 36), the upper novaculite member includes 192 feet of novaculite succeeded by 56 feet of subnovaculite, 155 feet of bedded brown chert, locally stained black by MnO_2, and finally by 25 feet of typical novaculite. The upper chert and shale is either absent or covered, because no trace of it can be found beneath a colluvial gravel layer.

As mentioned previously, red shale from 10 to 30 feet thick in the upper part of the upper chert and shale member is a distinctive unit in the central part of the basin.

Limestone beds are confined to the western part of the area, where they have been found at Localities 5 and 7. Sandstone beds above the lower novaculite have been found at Localities 2, 11, 14, 26, and 32.

Persimmon Gap and Rough Creek

Exposures of early Paleozoic rock occur at the eastern margin of the Marathon Basin at Rough Creek (Loc. 42) and between Persimmon Gap and Dog Canyon, about 10 miles south of the southernmost exposures of Paleozoic rock in the Marathon Basin.

The latter exposures are described by Maxwell (1949); Maxwell and others (1955, p. 54-56); Maxwell and Hazzard (1967, p. 23-28); Wilson (1954, p. 2462-2463); Hazzard and others (1958); and Flawn and others (1961, p. 62-63).

One exposure of interest is in the arroyo at Persimmon Gap, at the northern boundary of Big Bend National Park (Fig. 1). The exposed section includes about 100 feet of upper Maravillas Formation through the lower part of the Tesnus Formation. The rocks are shattered and folded beneath a Laramide thrust. Succeeding typical Maravillas black chert beds is a fissile, dark-olive-green shale, which is stained bright hematite red along joint faces and in spherical spots from 1 to 2 mm in diameter. The shale is jointed and breaks into domino-shaped pieces. A few beds of weathered and fractured light-green chert in beds from 1/16 to 3 inches thick are intercalated in the shale, mostly in the middle part of the unit. The shale is overlain by 94 feet of Caballos Formation, which is mostly dark-green to black chert and shale beds, but includes a novaculite member, 5 feet thick, that is 8 feet above the base (Fig. 8). The shale is deformed at its contacts with the Caballos and Maravillas cherts, and has been removed locally by thrusting.

The Rough Creek exposures have been described briefly by King (1937, p. 51), Wilson (1954, p. 2463) and Berry and Nielsen (1958), whose interpretations differ from each other and from my own. An area of several acres of Maravillas and Caballos outcrop is exposed on an anticline. About 100 feet of much fractured Maravillas is succeeded by, what appears to be, a complete (unfaulted) section of Caballos, which, in turn, is overlain by Tesnus. The Maravillas rests on Tesnus along a thrust fault.

Concerning the shale unit between the Maravillas and Caballos, Wilson wrote (1954, p. 2470):

> In all sections of older Paleozoic southwest of the Marathon region and in the Solitario a greenish and drab, light brown and pink weathering siliceous and gypsiferous shale is present between the Caballos novaculite and Maravillas chert. The writer proposes the name Persimmon Gap for this unit since the best section of easy access may be found $\frac{3}{4}$ mile up the gulch east of the road and ranger station at the Persimmon Gap entrance to Big Bend National Park. The unit is 25 feet thick here, 24 feet at Rough Creek, about 30 feet at old Jones Ranch, and slightly thicker, perhaps 40 feet, in sections in the Solitario. . . . The only fossils from the unit occur in concretions found in the basal 6 inches . . . in the Rough Creek section. The . . . fossils indicate Late Ordovician age.

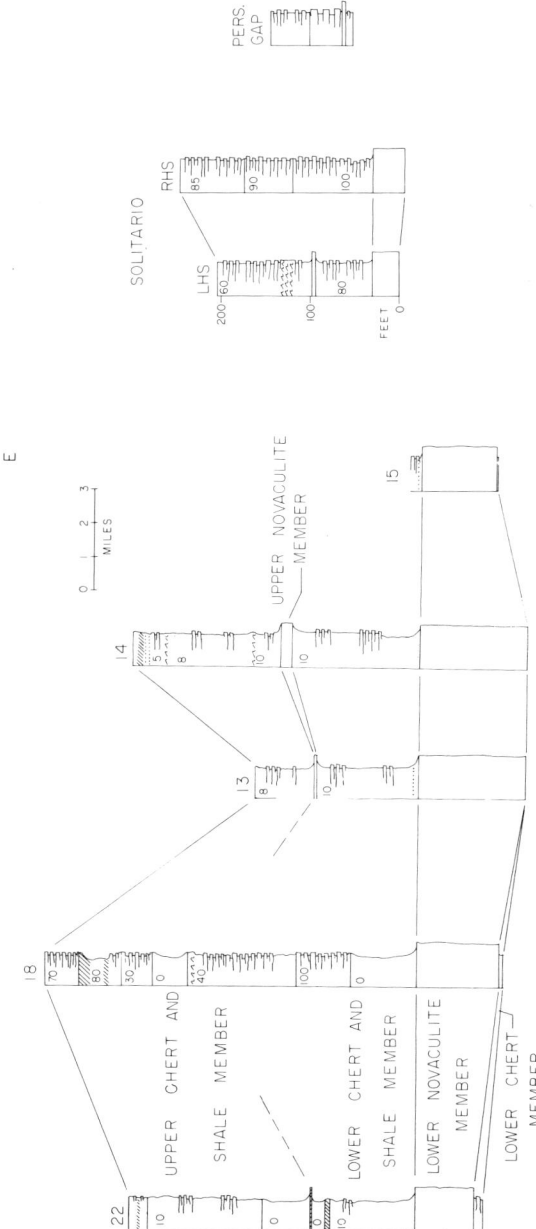

Figure 8. Stratigraphic cross sections of the Caballos Formation in the Marathon Basin, Solitario Uplift, and Persimmon Gap. See Figure 6 for explanation of symbols. Position of sections in Marathon Basin is along line C shown in Figure 3. Solitario sections are at Left Hand Shutup and Right Hand Shutup.

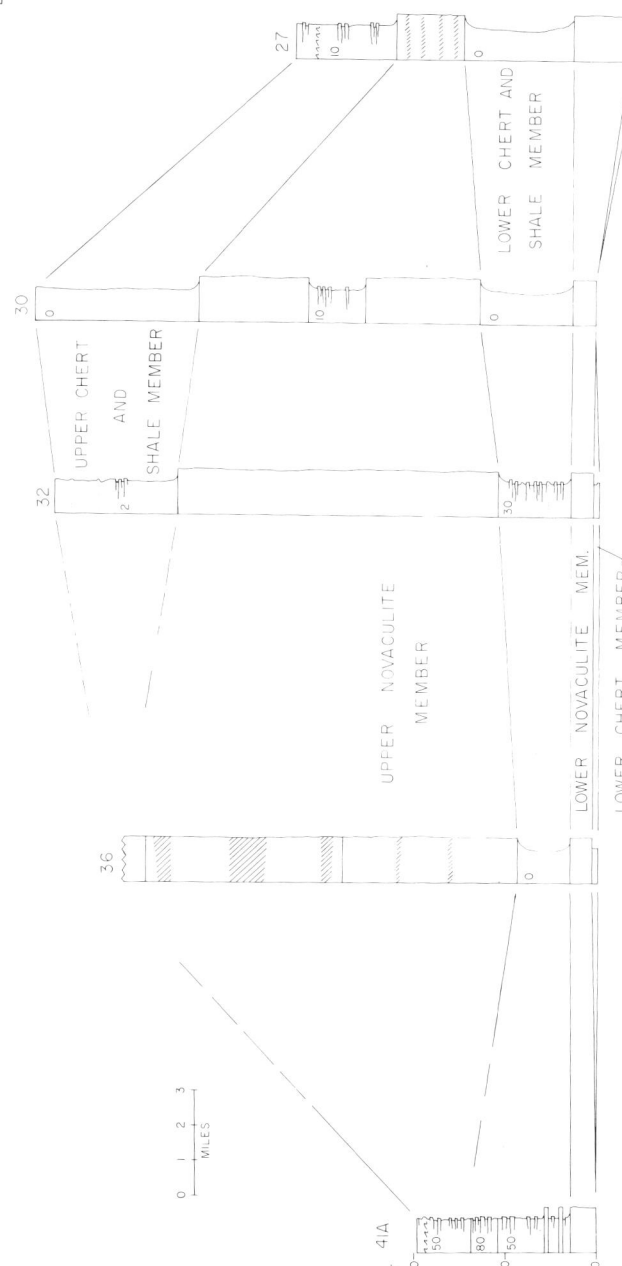

Figure 9. Stratigraphic cross section of the Caballos Formation in the Marathon Basin. See Figure 6 for explanation of symbols. Position of sections is along line D shown in Figure 3.

Berry and Nielsen (1958, p. 2259) believe the shale is part of the lower chert and shale member of the Caballos (my terminology) and that the fossils were in reworked concretions.

Unquestionably, a green shale unit occurs between typical Maravillas and Caballos lithologies south of the main Marathon outcrops from the old Jones Ranch to the Solitario uplift. I assign the shale under discussion to the upper Maravillas because such a shale, albeit with chert interbeds elsewhere, is present at the top of the Maravillas in the basin, and I believe the fossils at Rough Creek are in place.

The four interpretations made for the Rough Creek outcrop are shown schematically in Figure 10, where both Wilson's and my measured sections are generalized. Wilson measured 74.5 feet of Caballos, whereas I measured 89 feet; Berry and Nielsen are not specific on the thickness here. King (p. 51) reports 25 feet of bluish or white novaculite overlain by ". . . 175 feet of bluish, black or greenish dull lustered chert in 2 to 6 inch beds with thin partings of siliceous shale . . . " King did not mention Maravillas at this locality.

The shale unit beneath the Caballos is of unequal thickness around the limbs of the fold; it ranges from absent to 32 feet. Its position between two competent units makes it likely that it has been squeezed and boudined during folding, or possibly truncated by a thrust. If the latter is true, the Caballos that is exposed may be an incomplete sequence. The novaculite at Rough Creek is thinner than the closest novaculite in the basin to the west and thinner than the novaculite in the old Jones Ranch. Although part of the Caballos may have been removed by faulting at Rough Creek, it is equally likely that the thin section here is an original feature resulting from deposition on a submarine topographic high, an interpretation suggested by A. Thomson (1967, oral commun.).

Berry and Nielsen consider the novaculite exposed here to be the upper novaculite member. If the thin sequence of Caballos at Rough Creek (and Persimmon Gap) is equivalent in time to the thicker sections in the basin, an interpretation favored here, the novaculite at Rough Creek is equivalent in age to both novaculite and chert and shale members in the basin.

Old Jones Ranch

The Caballos and underlying shale that crop out on the old Asa Jones Ranch (presently McNutt lease of Slaughter ranch) in the Dove Mountain Quadrangle has been briefly mentioned by King (1937, p. 51), Wilson (1954, p. 2463-4), and Berry and Nielsen

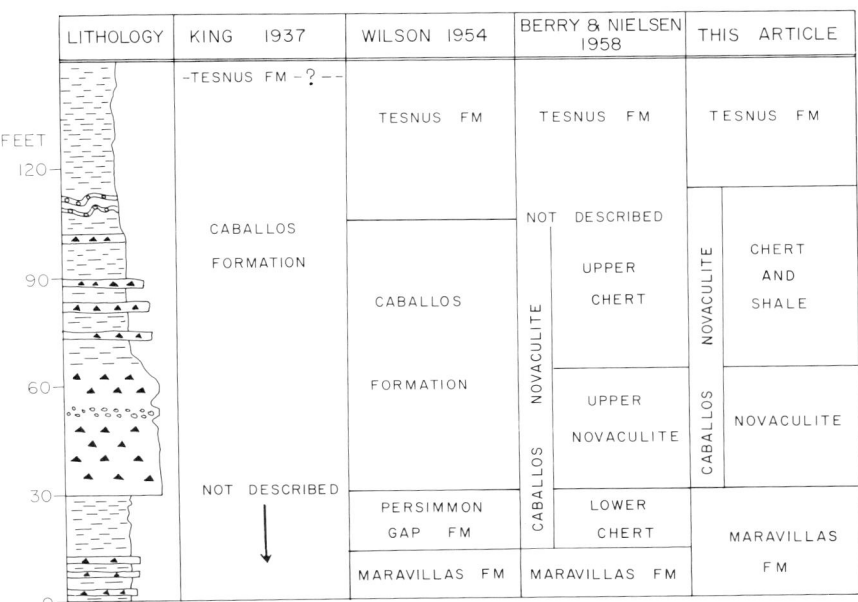

Figure 10. Stratigraphic section and terminology of Caballos and adjacent strata at Rough Creek (Loc. 42). Section is schematic; triangle symbol designates chert; contorted layers are slumped conglomerate beds.

(1958). When investigated during this study, access was made by the route described by Wilson.

The lower Paleozoic rocks are severely folded and faulted in this area, and it is doubtful whether accurate stratigraphic work can be done. At the western end of the pre-Tesnus outcrop, a small section of Caballos is exposed. The contact with the Maravillas is a fault in most places, and the Caballos outcrop appears to be largely thrust slices. In the least deformed section, about 60 feet of green siliceous shale in beds up to 3 inches thick is overlain by 175 feet of recognizable Caballos, which is succeeded by colluvium with float of Tesnus. The shale is Persimmon Gap according to Wilson, and middle Caballos according to Berry and Nielsen. I assign it to the uppermost Maravillas, following a similar procedure at Rough Creek and Persimmon Gap. Typical Maravillas black chert beds occur below the shale and in the shale 1 foot below distinctive Caballos lithology. The shale has at least one chert-pebble conglomerate bed up to 33 inches thick with clasts up to 3 inches long.

Of the 175 feet of Caballos exposed, I measured 48.5 feet of undisturbed section that includes off-white, blue-gray, and speckled subnovaculite chert with numerous shale partings (26.5 feet) followed by light-gray to black, blotchy, speckled, and mottled chert in beds from 2 to 6 inches thick with rare shale partings (22 feet). The following 70 feet is covered, except for a few gray chert beds, and the last 50 feet is a brecciated light-gray chert unit that rests with fault contact on the underlying beds. The latter could be a brecciated part of the lower subnovaculite and chert unit repeated by faulting.

On the same hill, but 100 yards to the east, is a sequence of bedded gray Caballos that is not exposed in the partial section described above. This sequence has a conglomerate bed 40 inches thick with slabs of Maravillas black chert and gray to brown-white (Caballos?) chert clasts up to 16 inches long in a light-brown chert host. The sequence rests with fault contact on Maravillas.

Berry and Nielsen (1958, p. 2257) believe the section here includes all three members of their Santiago Formation, or the upper three members of King's and my usage. The uncertain stratigraphy makes any correlation of Caballos units tenuous, but I differ in assigning the shale unit to the Maravillas rather than Caballos.

Solitario Uplift

Ridges of Caballos with the same general strike as in the Marathon Basin crop out in the Solitario Uplift. Near the Left Hand Shut Up, the formation is 218 feet thick; 5 miles to the southwest near the Right Hand Shut Up, it is 230 feet thick. Schematic stratigraphic sections are shown in Figure 8. The lower novaculite is 33 and 45 feet thick, respectively. It has many pods and kernels of blue-gray, tan, and gray chert. Bedding planes are pitted deeply by stylolites, and locally joints are stained by hematite and MnO_2. Beds of subnovaculite and blue-gray chert occur through 75 feet of section 60 feet above the novaculite, and may be correlative with the upper novaculite in the basin. Chert bedding planes in the upper part of the formation are commonly knobby. Several mottled, lumpy chert beds, deformed by submarine slumping and locally stained by MnO_2, are present in the upper part. Prospects have been staked in this MnO_2-rich unit.

The Caballos overlies approximately 45 feet of green fissile, slightly siliceous clay shale. The shale is stained red along bedding planes and joints, just as at Persimmon Gap, and has several interbedded chert layers from 0.5 to 2 inches thick.

PROBLEM OF THE LOWER CONTACT OF THE CABALLOS FORMATION

Two problems exist concerning the contact of the Caballos with the underlying Maravillas Formation: (1) the presence of an alleged unconformity, and (2) the position of the contact between the formations whether unconformable or not.

(1) Udden and others (1916, p. 41-42) and Baker and Bowman (1917, p. 93) reported the Maravillas to be overlain unconformably by the Caballos, and subsequent investigators, with the exception of Bennett (1959) and Thomson (1964) have accepted or agreed with this interpretation. Udden and Baker and Bowman's interpretation was based on the conclusion that Ordovician rock was overlain by Devonian rock. The Ordovician age of the Maravillas was based on fossils dated by Ulrich (*in* Baker and Bowman, 1917, p. 89) as Trenton (Fernvale-Richmond), whereas the Caballos was assumed to be the same age as the Arkansas novaculite, dated by Ulrich as early Devonian. Concerning the unconformity, Baker and Bowman stated (1917, p. 93):

> As the Maravillas is of variable thickness, it is probable that it is separated everywhere by an unconformity from the overlying Caballos novaculite. But such an unconformity between the two was actually observed only in the vicinity of Old Fort Peña.

Most subsequent workers in the area have endorsed the unconformity (King, 1930, p. 30; Sellards, 1933, p. 79; King, 1937, p. 21; Eifler, 1943, p. 1642; Graves, 1954, Pl. 8; Wilson, 1954, p. 2470). The endorsement stems largely from the fact that Silurian rock was presumably absent, and also because the uppermost beds of the Maravillas differ across the outcrop area, presumably due to removal of section by erosion. King (1937, p. 42) reports a conglomerate at the contact at several places, and implies that it is evidence of erosion.

Arick (1935, p. 120) first questioned the existence of the unconformity because "this contact, wherever seen, shows no evidence of conglomerate or erosion." Bennett (1959, p. 101-104) argued the contact was conformable in the area studied by him and that regional differences of uppermost Maravillas strata could be explained by facies differences equally as well as by erosion. The unconformity was denied also by Thomson (1964) and Thomson and McBride (1964), largely on the sedimentological interpretation that both the Caballos and upper Maravillas were deposited in deep water where sedimentation was continuous.

The present investigation supports both Bennett's and Thomson's conclusions stated above, and will be defended later.

(2) An additional problem concerns the position of the Maravillas-Caballos contact. The lithology at the contact was not described by Baker when the formations were named, and subsequent workers have placed the contact at different positions without being aware of a discrepancy. This matter was noted by Bennett (1959). An additional point is that the contact zone generally has undergone slippage and shear along bedding surfaces, because of differential competency of novaculite overlying shale and chert. Hence, it is unlikely that a truly undisturbed contact can be found anywhere in the region.

Throughout the Marathon Basin a shale-rich lithosome occurs between black chert and limestone of the Maravillas and off-white chert or novaculite of the Caballos Formation. The lithology varies laterally, but it is chiefly intercalated brown siliceous shale and chert in beds 0.5 to 4 inches thick. Measurements at 15 locations show that it ranges from 18 to 40 feet thick. (One value of 67 feet is based on the width of a topographic saddle between resistant black chert and white chert beds and may be an erroneous value.) The unit is poorly exposed and commonly forms a saddle between resistant Maravillas and Caballos chert, and is commonly covered by colluvial gravel from both formations. The unit was first obseved by Baker and Bowman (1917, p. 89) a few miles south of the town of Marathon. It correlates, in my opinion, lithologically and perhaps temporally with the less siliceous shale exposed beneath the Caballos at the old Jones Ranch, Rough Creek, Persimmon Gap, and Solitario Uplift exposures, and is synonymous with the Persimmon Gap Formation of Wilson (1954, p. 2470). The differences in thickness and lithology at these locations and in the Marathon Basin are considered to be facies changes of the uppermost Maravillas and not the result of post-Maravillas erosion. Such an interpretation of the Marathon Basin exposures was expressed by Bennett (1959, p. 46).

One of the best exposed and most referred to contacts of the Maravillas and Caballos is in the cliff above the picnic grounds at old Fort Peña (Loc. 13). Intermittent exposures of the contact visible over several hundred yards show considerable differences in lithology. In places the contact is clearly tectonic, because crumpled brown chert and shale (Maravillas) abuts fractured, but uncrumpled novaculite. Minor shearing between beds occurs where a chert-pebble conglomerate bed of discontinuous extent occurs beneath novaculite, whereas no discernible deformation occurs between beds where layers of brown siliceous shale and intercalated brown chert from 0.5 to 4 inches thick is immediately beneath novaculite (Pl. 6A). The brown chert and shale is several feet above typical black Maravillas chert. Berry (1960, p. 27) collected latest Ordovician graptolites from the unit and placed it in the Maravillas.

King's (1937) descriptions of the Maravillas and Caballos formations in general and of this particular outcrop do not make clear where he placed the contact. A measured section is given for the Maravillas, and a drawing (1937, Fig. 14) is given of the Caballos. King (1937, p. 42) refers to a "shale zone" in the Maravillas, but does not mention a shale-rich unit at the top of the formation in any of his Maravillas measured sections (p. 37-39). King's Figure 14 shows schematically several feet of lower chert member beneath the novaculite; elsewhere (p. 49), he described the lower chert member of the Caballos as " . . . brown, vitreous, splintery, thick-bedded chert." However, the lithic unit that I identify as lower chert member is not present at this locality. Assuming that we have chosen the same bed for the base of the novaculite member, and this is likely because the exposure is good, King's lower chert unit at this locality must be what I designate as the chert and shale member of the Maravillas Formation. The chert of this unit, however, is dull and thin-bedded. Thomson (1964, p. 15) referred to red and green cherts and shales of the lowermost Caballos Formation. He probably also had in mind the upper Maravillas chert and shale that weathers to a pinkish color, because red and green chert is not present in the lower Caballos as defined in this paper.

The upper brown chert and shale of the Maravillas is overlain by either a bed of white chert or novaculite that is here assigned to the Caballos. Discordant contacts between the Caballos and underlying beds that I have seen can best be attributed to structural deformation. Because of this deformation, and for two additional reasons, I believe that Maravillas-Caballos contact is conformable. The additional reasons are: (1) at several localities, white subnovaculite and light brown chert of Caballos and Maravillas lithology, respectively, are interbedded with 0.5 to 1 foot of section, and (2) I agree with Thomson's (1964) interpretation that both the upper Maravillas and lower Caballos are deep-water deposits between which an unconformity is improbable.

Exposures of intercalated shale of Maravillas lithology and

chert of the lower chert member of the Caballos are present at Localities 18 (East Bourland Mountain), 23, and along the Jeep road near the Right Hand Shut Up in the Solitario Uplift. The undisturbed? contact at the picnic grounds is abrupt, but conformable (Pl. 6A).

The conglomerate cited by King (1937, p. 42 and 50) as the basal bed of the Caballos, I place in the Maravillas. A conglomerate bed at this contact has been found at numerous locations in the central part of the basin. It is thickest (4 feet) at the picinc grounds, where locally it forms a pod of sedimentary slump origin 8 feet thick, and is 24 inches thick 6 miles east on the Granger anticline (Loc. 20), and is 14 inches thick 4 miles south (Loc. 18). The conglomerate has an intact framework of rounded pebbles of chert and chertified limestone and a matrix of chertified sand-size clasts. Pebbles of Maravillas-type chert are predominant, but white subnovaculite of Caballos-type chert is present. Pebbles are mostly less than 6 inches long, but slabs of chert up to 30 inches long have been found. Pebble-size clasts are not graded, but the top of the bed grades from granule to fine sand-size grains. At most localities the bed is a single sedimentation unit; possibly two sedimentation units are present at the picnic grounds. It is likely that the bed found at the same stratigraphic position at various localities is the same sedimentation unit (the product of one depositional event), but this cannot be proved. The presence of conglomerate at the contact has been cited as evidence in support of an unconformity. The pebbles certainly are evidence of erosion, but not necessarily at the site they are found. The pebbles include chert types different from the beds directly beneath the conglomerate bed; furthermore, the conglomerate was deposited on a soft substrate as shown by load-deformed and slumped contacts at its base. The conglomerate under question is interpreted as a submarine slide-turbidity current deposit (fluxoturbidite) emplaced in deep water. Conglomerates of this sort are common in many ancient deep-water deposits (Aubuoin, 1965).

PROBLEM OF THE UPPER CONTACT OF THE CABALLOS FORMATION

The upper contact of the Caballos has also been the subject of considerable discussion. Green chert beds 0.5 to 3 inches thick intercalate with gray-green siliceous shale beds half as thick in the upper Caballos. Siliceous shale continues upsection for several feet and gets progressively less siliceous and more fissile and passes into relatively soft, fissile, olive-drab clay-shale of typical Tesnus character. Beds of darker, resistant siliceous shale, a few inches thick, occur locally within the lower few tens of feet of Tesnus shale and alternate repeatedly with more fissile shale.

Distinct chertified beds of granule and fine-pebble conglomerate occur near the contact at many localities in the basin. The beds are of uneven thickness when traced along strike, and range from a few grains thick to 30 inches thick (Loc. 9). Variously colored chert particles of Caballos types, locally up to 3 inches long, are the chief clasts, generally in a quartz-rich sand matrix. About two-thirds of the beds have graded bedding.

Baker and Bowman (1917, p. 101) and King (1937, p. 52) describe the contact as being unconformable. King cited the following evidence: (1) Rocks of Mississippian age were thought to be missing; (2) the upper chert member of the Caballos is of unequal thickness; (3) conglomerate beds were considered evidence of erosion.

On the other hand, Bennett (1959, p. 111), Cotera (1962, p. 6), Thomson (1964), and McBride and Thomson (1964, p. 17) have interpreted the contact as gradational. King (p. 52) also noted that: "In some places, however, the boundary between the two formations is not easy to draw." Berry and Nielsen (1958, p. 2258) concluded that because the conglomerate beds are not continuous, no pronounced unconformity is present.

The pieces of evidence cited by King to document an unconformity can be differently interpreted; in addition, conodonts of Mississippian age have been found in the lower part of the Tesnus Formation (Ellison, 1962). The unequal thickness of the upper Caballos is inferred to be an original feature reflecting submarine topography. Thomson (1964, p. 15) argued that the conglomerate beds are deepwater deposits and not necessary evidence of *in situ* erosion. As noted by Bennett (1959, p. 111), most clasts in the conglomerate bed at the contact are different from the bed on which it rests; thus, the conglomerate heralds erosion at a location distant from its present site.

Because compelling evidence for an unconformity is lacking, the formations are considered conformable, and I support Bennett's and Thomson's views. The top of the Caballos was generally placed at the top of the last green chert bed that is only a few feet below fissile, olive-drab shale. Conglomerate beds occur locally below this contact, but are more common and thicker above this contact. Locally, the top of the Caballos was placed at the base of a prominent conglomerate bed where the bed occurs above the last chert bed, but below typical Tesnus shale.

FOSSILS AND AGE

For many years the Caballos Formation was assumed to be Devonian or questionably Devonian because of its similarity in stratigraphic position and lithology to the better dated Arkansas Novaculite (Baker and Bowman, 1917, p. 100-101; King, 1937, p. 52). The only closely dated fossils collected from the Caballos to date are conodonts from the upper chert members, reported at first by Graves (1952, p. 612) to be Late Devonian, and later by him (1954, p. 14) to be Middle to Late Devonian in age.

Radiolaria were recognized by Baker and Bowman (1917, p. 100-101), Henbest (1936, p. 77), Aberdeen (1940), Goldstein (1959, p. 148) and Thomson (1964). Aberdeen described 18 genera of *Spumelaria* that were found in samples of green chert of the "Santiago member," collected by Baker. Thomson (1964) found Radiolaria and siliceous sponge spicules throughout the formation, an occurrence substantiated by this study.

King (1937, p. 52) reported linguloid brachiopods in a limestone lens 1 mile west of the Roberts ranch.

Bennett (1959, p. 78-92) found numerous pieces of petrified wood (*Callixylon* cf. *newberryi*) of Devonian age in the upper Caballos (his members 5 and 6) on East Bourland and Simpson Springs Mountains. The largest piece is a remarkable log 28 feet long and 14 inches in diameter. Additional fragments of logs have subsequently been found south and east of the above area. Bennett also reports the presence of fecal pellets and the following questionable fossil forms: burrow mounds and pits, gastropod, pelecypod, and plant remains. Photographs of the gastropod and pelecypod remains are unconvincing to me.

Goldstein (1959, p. 145-6) describes the presence of traces of phosphatic or chitinous fossil fragments, or both, and spore exines.

The following fossils were found during the present study: Radiolaria, sponge spicules, conodonts, wood, pits, mounds and trails and burrows made by benthonic animals, plates of phosphate or chitin of unknown affinity and calcite skeletal grains of echinnoderms, brachiopods, blue-green algae, trilobites(?), and glauconite and silicified fecal pellets. Electron microscopy has revealed the presence of stellate and circular objects (Pl. 9B) of uncertain affinity, but resembling spicules and recrystallized Radiolaria. The calcite skeletal fragments are from calcarenite limestone beds in the Payne Hills, from which the dated conodonts have been collected. Table 1 shows the vertical distribution of the faunal elements.

The phosphate or chitin fragments mentioned by Goldstein (1959, p. 145-6) are from sandstone at the base of member 2; additional fragments occur in calcarenite. My samples show they are elongate, probably platy objects with a maximum thickness of 0.3 mm and length of 1.2 mm. Only one particle has a scalloped edge, the remainder are smooth-edged. Light and dark bands run the length of the objects, and in a few give the impression of an axial core. King (1937, p. 52) found linguloid brachiopods in the calcarenite; hence, the particles probably are small fragments of this taxon.

The bulk of the formation lacks datable fossils. The limestone beds, from which Graves collected Middle and Late Devonian conodonts, are in the lower chert and shale member. The closest dated horizons to this member are the Late Ordovician Maravillas graptolites immediately below the lower Caballos chert, and Late Mississippian conodonts 30 feet above the base of the Tesnus Formation.

Apparently Jones (1953, p. 17) suggested first that the formation may range from Early Silurian to Early Mississippian in age, and Bennett (1959, p. 104) suggested that the Caballos might also include rocks representing the time interval between latest Ordovician and Late Devonian.

Thomson and McBride (1964) previously have supported the interpretation that sedimentation was continuous in the axial portion of the Marathon geosyncline from Late Cambrian through Late Pennsylvanian time. Continuous sedimentation across the Caballos contacts was argued also specifically by Thomson (1964). According to our interpretation, the Caballos can be post-Late Ordovician to pre-Late Mississippian in age; it certainly includes rock of all of Silurian time, most of Devonian time, and perhaps rock of Early Mississippian and Late Ordovician time. The absence of fossils dated as Silurian or Early Devonian age is damaging to the above interpretation. However, this defect is not so significant when one considers the small number of taxa recovered from the formation and the poor degree of preservation of most individuals.

TABLE 1. STRATIGRAPHIC DISTRIBUTION OF FAUNAL ELEMENTS

	Radio-larians	Sponge spicules	Trace fossils	Cono-donts	Petri-fied wood	Phos-phate or chitin plates	Echino-derms	Brachio-pods	Algae	Trilo-bites(?)	Glauco-nite fecal(?) pellets
UPPER CHERT & SHALE	X	X	X	X	X						
UPPER NOVA-CULITE	X	X	X								
LOWER CHERT & SHALE	X	X	X	X		X	X	X	X	X	X
LOWER NOVA-CULITE	X	X	X								
LOWER CHERT	X	X	X								

DISCUSSION

(1) The original shape of the Caballos Formation is unknown. However, the Caballos lithosome probably extended continuously from the Solitario Uplift to the Marathon Basin. In the Marathon Basin, Caballos sediments accumulated to form a lens-shaped unit that, when viewed palinspastically, is elongate northwest-southeast and is normal to the present structural grain. The Marathon sedimentary basin may have been a more rapidly subsiding part of the linear Ouachita geosyncline or an embayment off the main trough. Daly (1964) promoted the latter interpretation, and thought the Marathon Basin was an exogeosyncline that postdated the filling of the main linear orthogeosyncline. Data are yet too meager to document Daly's interpretation.

(2) Regional relations of the Caballos and correlative rocks are poorly known because rocks of equivalent age do not crop out in surrounding areas, and subsurface information is spotty, except in the Midland Basin to the north. The picture is complicated by the uncertainty of how far the present outcrops have moved from their original position during Pennsylvanian thrusting. The Gulf Oil Company's No. 1 D. S. Combs and the Slick-Urschel Oil Company's No. 1 Decie Sinclair wells in the western part of the Marathon Basin penetrated rocks of the foreland (cratonic) facies beneath deformed rocks of the Marathon (or Ouachita) facies (Wilson, 1954; Flawn and others, 1961, p. 234-238; Ross, 1962). Hence, the Marathon facies is clearly allochthonous along the western part of the basin. Wells to the east of the basin do not pass through rocks of the Marathon sequence.

Conditions of deposition were not uniform between the Marathon Basin and the Oklahoma-Arkansas part of the Ouachita geosyncline, because only a dark chert facies has been recovered from deep wells between the areas (Flawn and others, 1961).

(3) A regional subsurface study of Siluro-Devonian rocks in west Texas by McGlasson (1967) suggests the Caballos is correlative to four lithologic units that were deposited in the Tabosa Basin, a depositional basin which occupied west Texas and southeastern New Mexico during early Paleozoic time and which opened southward into the Marathon geosyncline. The correlative units include: (a) the Fusselman Formation, (b) an unnamed "Upper Silurian" unit, (c) an unnamed "Devonian" unit, and (d) the Woodford Formation. The Fusselman Formation (Silurian) is limestone and dolomite; the "Upper Silurian" unit is shale, limestone, and

dolomite; the "Devonian" unit is chert and siliceous limestone; and the Woodford Formation is dark shale. Correlation with the Caballos is based in part on lithology and in part on the age interpretation proposed herein.

(4) Evidence suggests that the upper and lower contacts of the Caballos are nearly isochronous surfaces, at least in the central part of the Marathon Basin. Evidence at the lower contact is the occurrence of latest Ordovician graptolites (Berry, 1960), a few inches below the contact at many localities, and the presence, at the contact, of a conglomerate bed interpreted to be a slump deposit or turbidite of isochronous character. Evidence at the upper contact is less conclusive, but contorted chert beds that record instantaneous slump events are common within 50 to 100 feet of the contact. Because many localities have more than one slumped bed, individual beds cannot be precisely correlated among adjacent sections, however.

(5) An important question to a time-stratigraphic analysis is whether each novaculite member represents the same interval of time throughout its extent (Fig. 11A); that is, whether each has isochronous boundaries. Such an interpretation was implied by Berry and Nielson (1958). In Part 2 of this report, it is argued that the distinct characteristics of novaculite arose from a combination of depositional and diagenetic events. Hence, in the absence of good faunal control, the novaculite members cannot be proved to be time-stratigraphic units. Novaculite intertongues with bedded chert and shale in the southern limb of the Dagger Flat anticlinorium and shows that some novaculite beds formed contemporaneously with colored chert beds. However, the generally abrupt lithologic change at the boundaries of the novaculite members in the central part of the Marathon Basin suggests that the members are time-stratigraphic units in that area (situation A of Fig. 11); possibly the members are time-transgressive (situation B of Fig. 11) at the basin margins.

(6) The Caballos Novaculite generally has been correlated with the Arkansas Novaculite of the Ouachita Mountains on the basis of similar lithology and similar position in the geosynclinal section. However, if the Caballos ranges in age from Silurian through Devonian and possibly Early Mississippian time, as maintained herein, it is equivalent in age to the Arkansas Novaculite plus the Blaylock Sandstone and Missouri Mountain Shale;

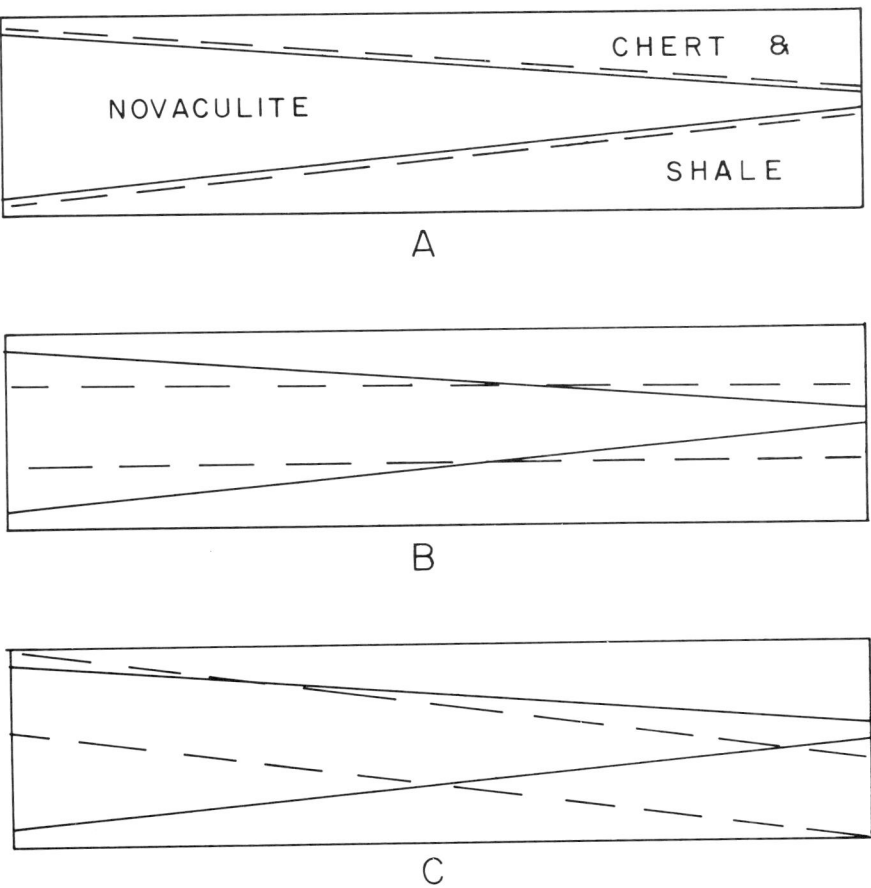

Figure 11. Diagrammatic cross section showing possible relations of time lines to novaculite and chert and shale members. Dashed lines are isochrons.

the latter formations are of Silurian age (Sterling and others, 1966). The relations of the three members of the Arkansas Novaculite to the five members of the Caballos Formation must remain speculative until adequate biostratigraphic control is available.

(7) In a discussion of Paleozoic sequences (lithic units bounded by interregional unconformities; Sloss, 1963) Wheeler (1963, p. 1511) assigned the lower part of the Arkansas Novaculite to the uppermost part of his Tutelo Sequence, a subdivision of

the Tippecanoe Sequence described by Sloss (1963, p. 97-99). Sequence boundaries are not present in the Marathon region. As interpreted here, the Caballos spans the boundary of the Tippecanoe and Kaskaskia Sequences. This supports Sloss's views that major unconformities tend to disappear at the cratonic margins into continuous successions.

Part 2. Petrography and Origin

EARLE F. MCBRIDE
and
ALAN THOMSON

Petrography and Origin

INTRODUCTION

Part 2 of this report concerns the petrography and petrogenesis of the bedded chert and associated beds of the Caballos Novaculite. In addition to conventional field work, more than 250 thin sections were studied. Selected samples were analyzed by X-ray diffraction, and chemical analyses were made of 13 samples. Replicas of several types of chert were studied using the electron microscope. The locations of samples referred to are shown in Figure 3.

Our work on the Caballos began independently several years ago. The academic position of McBride permitted summers for study; most of the petrographic work was done by him, and the manuscript was written by him. From repeated discussions our ideas on origin have been traded and intermingled such that the responsibility for interpretation is shared by both equally. Summaries of the interpretations presented here appeared previously in field guidebooks (Thomson, 1964; Thomson and McBride, 1964; McBride and Thomson, in press) and in abstracts (Thomson, 1965; McBride, 1968b).

PETROGRAPHY

General

Not only do Caballos chert beds differ considerably in color, but they also show wide variations in grain size, fabric, and amount of miscellaneous clastic and authigenic constituents. For this part of the report, rocks of similar megascopic character are described collectively. Many rocks, such as green chert and green siliceous shale, are completely gradational in composition, and the boundary between the types has been decided arbitrarily.

There has been little uniformity in the use of textural and mineralogical terms applied to cherty sedimentary rocks because of their fine grain size and, commonly, uncertain mineralogy (see Pettijohn, 1957, p. 432-438). Hence, petrographic terms used in this report are defined below for clarity. Several terms are modified from the definitions proposed by Folk (1965, p. 80), and Folk and Weaver (1952). Grain-size values are those measured in thin section.

CHALCEDONY: quartz with a radiating fibrous structure.

CHERT: a sedimentary rock composed predominantly of microquartz and small amounts of megaquartz and other impurities.

MEGAQUARTZ: quartz grain larger than 35 μ.

MICROQUARTZ: quartz grains smaller than 35 μ. Microquartz is here subdivided into the following grain-size classes:

COARSE GRAINED (CG) = 25 to 35 μ.
MEDIUM GRAINED (MG) = 15 to 25 μ.
FINE GRAINED (FG) = 10 to 15 μ.
VERY FINE GRAINED (VFG) = 5 to 10 μ.
ULTRA-FINE GRAINED (UFG) = less than 5 μ. UFG microquartz is cryptocrystalline.

All varieties of microquartz show undulose extinction in thin sections of normal thickness. The undulosity is a net effect resulting when light that passes through any point on a section penetrates several small grains of uneven thickness and different crystallographic orientation (Folk and Weaver, 1952, p. 500; Carver, 1965).

NOVACULITE: chert that has a milk-white color in hand specimen.

SUBNOVACULITE: white and off-white chert that is not milky.

Novaculite

Novaculite fragments of hand-specimen size range from milk-white, featureless, homogeneous rocks (Pl. 1) devoid of bedding to those in which bedding is shown by off-white color bands, shale partings, or by rare, smooth bedding planes. Most conspicuous bedding planes have been altered to prominent stylolites (Pl. 5B). Various degrees of milkiness (Pl. 1) are displayed, and all samples are translucent in all but thin slices. Fractures from hairlines to 5 mm wide that are healed by megaquartz, and isolated ellipses (eyes) of megaquartz up to 5 mm in diameter are prominent in some beds. Commonly fracture surfaces, which are semiconchoidal, are unevenly coated by red hematite or MnO_2. Pinpoint spots of probable recrystallized Radiolaria and sponge spicules are discernible with a hand lens in a few beds. The spicules commonly are aligned so that they define bedding.

Thin sections show (Pls. 7A, 7B, 8A, and 14D) that novaculite is composed of 90 to 99 percent microquartz, a trace of 4 percent detrital and authigenic grains, and variable amounts of megaquartz in healed fractures and eyes. Novaculite has a distinct internal texture that is different from other varieties of chert in the formation. The *novaculite texture,* discernible under crossed polarizers only, is a slightly uneven-grained groundmass predominantly of MG microquartz, in which subspherical specks of a mixture of UFG and FG microquartz are scattered. Microquartz grains of sizes other than MG occur in the groundmass in amounts up to 20 percent. FG and CG microquartz or mixtures of MG to FG and MG to CG microquartz form the groundmass of a few samples. Chalcedony is absent in most samples, but is present from a trace to 20 percent in 8 of 25 samples studied. Chalcedony is distributed randomly in the groundmass, except in the few samples where it fills radiolarian capsules. Abundances of the major components of the 25 samples are given in Table 2.

The specks are mostly slightly elliptical or subspherical objects (Pls. 7B and 7C) that range in size from 10 to 200 μ. They comprise from 3 to 60 percent, but rarely exceed 30 percent of a sample. Their distribution is generally random, although in some samples layers occur that have greater or lesser than normal density. Most specks have poorly defined ragged boundaries and grade into the groundmass by gradually increasing grain size. Specks are invisible in plain light in some slides, but have a pale- to medium-brown color in others. The brown-colored part ranges from a trace to 60 percent of a rock, but rarely exceeds 30 percent.

Bedding planes within samples are absent, except for rare occurrences of laminae of quartz silt less than 1 mm thick. The

TABLE 2. COMPOSITION AND TEXTURE OF CABALLOS NOVACULITE SAMPLES
Values in Volume Percent of Total Rock, Tr = 1 or 2 grains per thin section

Sample no.	Quartz	Radiolaria ghosts	Spicule ghosts	Specks	Chalcedony	Bubble inclusions	Grain Size of matrix	Brown cast	Comments
21	1	0	Tr	5	0	30 Prominent 10 Faint	FG-MG	20, local; ramose inclusions	Mottled, no good specks
23	Tr	Tr	Tr	8	0	10 Prominent 10 Faint	MG	80, faint	Pellets
24	1	1	Tr	8	Tr	5 Prominent 5 Light	MG	60, faint, pebbly	Specks poorly defined
68	1	1	Tr	60	0	25 Dark 10 Light	FG	60, checkerboard prominent	Prominent bubbles
69	3	0	Tr	35	0	Sparse	FG	40, checkerboard	Specks poorly defined
74	1	2	5	5	0	Sparse	MG	5, checkerboard at joints	Specks poorly defined
76	Tr	1	1	30	0	5 Prominent Tr Faint	MG	10, checkerboard some Mn-oxide	Specks like tapioca, huge bubbles
77	Tr	25	2	25	0		MG	local ramose inclusions	Specks clearly are Radiolaria
119	Tr	0	10	10	0	Sparse	FG-MG	30, faint	Megaquartz eyes, fecal pellets
120	1	0	50	8	8	Tr	MG	Tr	Excellent ghosts
121	Tr	0	Tr	10	3	Tr	FG-MG	Tr	

TABLE 2. (Continued)

Sample no.	Quartz	Radio-laria ghosts	Spicule ghosts	Specks	Chal-cedony	Bubble inclusions	Grain Size of matrix	Brown cast	Comments
123	1	Tr	5	8	Tr	10 Faint Tr Prominent	MG	Tr	
131	Tr	0	10	15	0	Tr	MG-CG	10, faint	
133	Tr	0	5	5	15	Tr	MG-CG	Tr	
142	3	1	2	10	0	1	MG	30, locally strong	Burrowed, good pellets
146	1	0	40	7	Tr	Sparse	FG-MG	3, faint	Some pellet specks
151	1	0	6	3	0	Tr	FG-MG	5, faint	Pyrite inclusions
171	Tr	0	3	25	0	Tr	MG	Tr	
P1	1	Tr	1	35	0	Tr	MG	10, faint	Some megaquartz
P2	Tr	Tr	30	10	0	Tr	MG	3, faint	5 megaquartz, Tr pellets
P3	1	0	80	20	3	5 Faint	MG-CG	5, very faint	Excellent ghosts
CP1	Tr	Tr	8	15	20	15 Faint	MG	50, dark locally	2 megaquartz, some recrystallization
X54	1	Tr	5	20		10 Faint	MG-CG	10, prominent	Spicules replaced by phosphate (?)
Sr-1	1	0	0	5	0	30 Prominent	FG-MG	70, prominent	Burrow nest
SR-9	Tr	0	1	20	0	10 Faint	FG-MG	10, in specks	3 spherical specks

direction of bedding is always discernible, however, by the direction of flattening of the specks or the trend of speck-rich layers.

Common impurities in novaculite include detrital quartz and authigenic pyrite, hematite, and well-formed rhombs of manganocalcite. Much hematite is pseudomorphous after pyrite euhedra. Quartz grains of silt and locally of very fine sand (up to 0.08 mm in diameter) are present from one grain per slide to 3 percent of a sample; the other impurities are present as scattered grains, generally in only trace amounts. The detrital quartz grains are commonly rimmed by a thin, unabraded secondary overgrowth, and are randomly scattered (Pl. 14D), except for the rare laminae mentioned previously. Illite particles occur in 2 to 3 samples where shale partings were mixed into novaculite by burrowing animals. MnO_2 occurs in a few areas in novaculite beds that have closely spaced fractures adjacent to faults. X-ray diffraction patterns of the less than 2 μ fraction of powdered novaculite on oriented slides show only the presence of well-crystallized quartz.

Rhombs of a carbonate mineral (Pl. 11B) occur in all chert varieties in the formation, but are least common in novaculite. The rhombs are more spindle-shaped than typical dolomite rhombs and appear to have both indices in refraction above Canada balsam. Most grains alter to an opaque material that has a dull luster characteristic of pyrolusite; hence, the carbonate mineral is inferred to be manganiferous calcite. Grains similar to those in the Arkansas novaculite were identified as rhodochrosite by Honess (1923, p. 133-4).

Thin sections of novaculite samples viewed in ordinary light under low-power magnification range from clear, transparent slides to those with blotchy shades of pale brown; locally conspicuous hematite-coated fractures are fairly common. Interesting details can be seen in many seemingly featureless novaculite thin sections when the microscope is slowly focused above and below the plane of best focus in ordinary light. Becke lines are visible between microquartz grains, whose indices of refraction differ either because of different orientations or water-bubble content. Becke lines outline curved, ghostlike shapes composed of several grains of microquartz. Most ghost shapes have no distinct patterns, but some are sections of sponge spicules (Pl. 8A) and Radiolaria. Spicules are elongate tapered monaxon forms up to 0.6 mm long and 0.08 mm wide that have a central canal that is slightly greater in diameter than the wall thickness. Jointed spicules are rare.

Particles identified as Radiolaria have circular cross sections, with a maximum diameter of 0.18 mm, that are composed of pale brown UFG and VFG microquartz devoid of internal structure. These are probably internal molds of Radiolaria. The percentage of novaculite that is composed of recognizable spicule or Radiolarian ghosts ranges from a trace to 80, but is generally less than 8. In most samples studied, spicules are present nearly to the exclusion of Radiolaria. Novaculite surfaces etched by hydrofluoric acid clearly show the skeletal components in reflected light (Pl. 8B), even where no details were visible before etching. However, etching did not reveal organisms in rocks in which ghosts are not visible in thin section.

Many samples contain conspicuous small clusters of inclusions that are dark brown in transmitted light and silvery in reflected light. These inclusions are most common in specks of microquartz or along healed fractures, but locally pervade large areas of a sample. Although conspicuous, they probably make up only a few percent of the rock by volume. Magnification of 1000X shows the inclusions are spheres or ramifying cylindrical tubes about one micron in length and have high negative relief. Possibly the inclusions are water-filled cavities similar to those in chert samples studied by Folk and Weaver (1952, p. 501), where the brown color was attributed to a dispersion effect. Although a few bubble inclusions are visible in the pale-brown specks in novaculite, there is no other evidence for the cause of the brown color of the specks.

Some specks of UFG and VFG microquartz are clearly the remains of recrystallized Radiolaria, a few are ghosts of sponge spicules, and some are probably fecal pellets. Specks in several samples (Pl. 7C) are circular in cross section and have a well-defined outer wall; these are certainly Radiolaria. The same rock contains all gradations from these large specks to small ones that have fuzzy boundaries and only crudely circular form (Pl. 7B). Other samples have specks that are elliptical and that lack well-defined boundaries, and a few have inclusions of randomly oriented clay particles and blebs of hematite. Whereas specks of Radiolarian origin do not exceed 0.2 mm in diameter, specks of probable fecal origin range from 0.05 to 0.6 mm. Specks that convincingly may be attributed to either a fecal or Radiolarian parent comprise only 1 to 2 percent of the samples studied; however, the remainder are presumed to be of similar origin. The specks are clearly different from the organic structures described in Precambrian iron formations (Barghoorn and Tyler, 1965; LaBerge, 1967) because they are larger and internally structureless.

Chemical analyses (Table 3) were made of five samples of novaculite. The silica content exceeds 98.9 percent, and most other oxides are present in amounts less than 0.1 percent. Al_2O_3 from illite is the chief impurity.

Studies of novaculite surface replicas using the electron microscope have been made by Folk and Weaver (1952), Park (1961), and Park and Croneis (1969). Folk and Weaver studied several samples of Arkansas novaculite as part of a general study of chert. Samples studied by them were described as homogeneous chert composed largely of polyhedral blocks of microcrystalline quartz; they designated surfaces composed of these blocks as having a "novaculite type surface." Park (1961) studied 22 replica surfaces of different chert types from the Caballos and Arkansas Novaculite Formations, but did not specify the number of different beds they include. Park (p. 36) states that

... surfaces from the novaculite formations ... illustrate the typical novaculite type morphology made up mostly of blocks smaller than three microns and generally between one and one-half microns, although there are areas which are made of blocks which are smaller. The smaller blocks appear to be interspersed with the large blocks in veinlets or narrow bands.

For the present study, replicated surfaces were made of five samples of novaculite, one subnovaculite, two green chert, and one blue-white chert. Platinum-shadowed carbon replicas were made of both sawed surfaces and fracture surfaces that were etched by HF. The latter preparation yielded the best results, although there was considerable variation in the quality of the replicas.

The five novaculite samples selected for study differ noticeably among themselves in textural details visible in thin section, and are equally different in electron micrographs. The latter show that four of the five samples are characterized by a distinct bimodal grain size and also have sparse patterned areas that are probably relict skeletal remains (Pls. 9B and 10A). The fine mode is grains between 0.1 and 0.5 μ, whereas the coarse mode is between 3 and 6 μ. Most grains are polygonal with slightly curved boundaries (Pl. 10B); the coarse grains are equant, whereas the fine grains are commonly elongate by a ratio of 2:1. Irregular grains with interlocking jigsaw-puzzle fabric are present but uncommon. Details of each sample studied are summarized in Table 4.

TABLE 3. CHEMICAL ANALYSES OF CABALLOS NOVACULITE SAMPLES

Numerals directly below the rock names are sample numbers.
No determination was made of constituents shown as dots.

	Novaculite						Blue-white chert	Green-black chert	Green-gray chert	Dull-green chert	Cherti-fied congl.	Red shale	Green shale	Cal-care-nite
	1	2	113	138	68		111	38	164	139	79	32	33	162
SiO_2	99.60	98.90	99.20	98.90	99.40		98.65	96.10	93.00	87.40	93.10	59.60	65.82	11.36
TiO_2	.00	.01	.01	.02	.00		.02	.07	.16	.22	.02	.74	.66	.02
Al_2O_3	.06	.30	.24	.22	.15		.29	1.00	2.80	4.90	.32	14.30	14.30	.06
Fe_2O_3	.10	.01	.02	.01	.03		.02	.43	.26	1.21	.71	8.91	.70	.19
FeO	...	...	...	...	...		...	.10	.10	.24	.25	.79	4.02	.14
MnO	.00	.01	.01	.01	.00		.01	.01	.02	.01	.03	.00	.00	.06
MgO	.02	.02	.03	.02	.03		.02	.09	.47	1.00	.35	2.00	2.00	.54
CaO	.01	.04	.07	.14	.08		.09	.07	.08	.13	.79	.18	.75	48.00
Na_2O	.09	.04	.05	.10	.04		.06	.08	.10	.06	.13	.62	.60	.30
K_2O	.02	.02	.02	.03	.02		.04	.30	.48	1.70	.51	4.80	4.50	.03
*H_2O+	.08	.13	.22	.18	.18		.20	.55	.91	1.32	.73	3.40	2.80	.15
H_2O-	.06	.08	.02	.08	.14		.08	.31	.53	.45	.16	2.45	2.15	.15
*CO_2	.09	.19	.15	.16	.16		.22	.15	.25	.09	1.80	.13	.12	...
P_2O_5	.00	.00	.004	.01	.17		.005	.008	.005	.008	.09	.04	.03	.09
SO_3	.13	...	...	...	...		...	...	...	.08	.70	.02	...	...
Ign. loss	.22	.37	.26	.53	.28		.66	1.09	1.56	1.92	2.13	4.37	4.32	38.48
Total	100.31	99.80	99.93	100.07	100.34		99.94	99.65	99.56	99.33	99.29	98.82	99.85	99.42

*Value is excluded from the total because it is included in the ignition loss.

TABLE 4.

DESCRIPTION OF CHERT TEXTURES IN ELECTRON PHOTOMICROGRAPHS

Novaculite Sample

68. Distinct bimodal grain size (Pl. 9A). Fine mode = 0.5-1.5 μ; coarse mode = 3-6 μ. Patches of smaller grain size are circular, but have no internal structure. The boundary between fine and coarse is either sharp or has a narrow transition zone. Both coarse and fine particles have crude subequant polygonal form (Pl. 10B). Slightly curved boundaries are most common, but strongly curved boundaries are present. Surfaces of larger grains are locally marked by triangular pits presumably produced during etching. Pits are uniformly oriented and demonstrate crystallographic control.

76. Low magnification shows circular areas approximately 6 μ in diameter that have very crude radial pattern; circles comprise 10 percent of the field. Coarse grains of matrix = 3 μ; fine grains of circular patches = 0.3 μ. Spokes of radials are locally well defined, but are devoid of internal structure.

142. Chiefly large, equant grains (3.0-5.0 μ) with straight to slightly curved boundaries and with local surface ornamentation of parallel lines or triangular pits. A mosaic of finer-grained particles (0.1-0.5 μ) that are more circular than polygonal occur between approximately one-third of the larger grain contacts. A few spheres (water bubbles?) occur among the mosaics.

P-3. Distinct bimodal texture. The fine mode is grains with slightly curved boundaries (0.2-0.4 μ); the coarse mode (3.0-8.0 μ) is grains with straight boundaries (Pl. 10B) and prominent crystal faces that are locally textured by curved parallel lines. Sparsely scattered through the rock are six-pointed star-shaped objects (3 μ in diameter) with arms that join at angles at 120° (Pl. 9B). The stars are composed of uneven-grained irregular-shaped particles approximately 0.1 μ in diameter.

ZPG. Chiefly composed of even-grained particles from 0.1-0.3 μ in diameter. About 15 percent of the low-magnification field (3 sq μ) is made up of circular areas with a radial pattern of particles (Pl. 10A). The spokelike radials are composed of small particles similar in size to the matrix, but hooked up chain-fashion and with indistinct boundaries between particles. Some circular patches are in contact, but others are separated by small grains of blocky quartz.

Subnovaculite

C-136. Chiefly equant grains from 0.7-2.0 μ in diameter; about one-third are twice as long as wide. Grains are polygonal with slightly curved sides.

Green Chert

36. Fairly even-grained (0.3-0.6 μ) rock that contains 10 percent large elongate grains up to 20 μ long that are probably silicate minerals. Long prismatic needles (0.2-3.0 μ) are a minor constituent.

37. Fairly even-grained (0.7-2.0 μ) rock with irregular grain boundaries. One field shows a paisley-shaped object 23μ long that is composed of doubly terminated bipyramidal crystals. The object is outlined by quartz grains coarser than the matrix.

Blue-white Chert

111. Prominent bimodal texture: coarse mode (3.0-5.0 μ) is equant grains with straight to slightly curved boundaries; fine mode (0.3-0.6 μ) is more irregular-shaped grains. One photo shows a myriad of poorly defined spherical objects (0.2 μ in diameter) that are either bubbles or artifacts.

Samples 76 and ZPG have circular patches from 6 to 10 μ in diameter that are composed of fine-grained particles. The patches in sample ZPG have a distinct radial fabric formed by elongate particles hooked up in a spokelike arrangement (Pl. 10A); this fabric is poorly shown in sample 76. The circular patches locally abut other patches, but elsewhere are separated by finer or coarser particles. In the thin sections of these rocks, faint ghosts of organisms can be seen locally in the coarse-grained particles (MG microquartz of thin-section designation), but the clots of UFG and VFG microquartz are structureless. The fine-grained patches in the electron micrographs undoubtedly correspond to the clots of UFG and VFG microquartz visible in thin section, although the estimates of particle size in thin section are up to 10 times larger than the size of particles seen in electron micrographs.

The radial fabric of the circular patches in samples 76 and ZPG probably developed upon the crystallization of amorphous opaline skeletal grains to microquartz. The presence of only circular patches in the electron micrographs suggests the skeletal grains were Radiolaria or spherical nannoplankton rather than sponge spicules.

Sample P-3 has the best preserved ghosts of skeletal grains (donut- and rod-shaped forms) of all novaculite samples studied (Pl. 8A, 8B) in thin section, but they are not recognizable on the electron micrographs. The sample is unique, however, in that it contains six-pointed star-shaped objects (Pl. 9B) 3 μ in diameter that comprise less than 1 percent by area of the replicas studied of this sample. The stars are composed of uneven-grained irregular-shaped particles approximately 0.1 μ in diameter.

The stars resemble some hexact megasclere spicules (de Laubenfels, 1955), but are much smaller. The affinity of this taxon of probable nannoplankton is unknown. They are the only ultramicroscopic taxon found to be so well preserved in the formation.

The novaculite surfaces studied are similar in most details to those described and illustrated by Park and Croneis (1969), except for the fabric of the circular patches and stars noted above. Grain boundaries of Park and Croneis's and our samples are more irregular and curved than those illustrated by Folk and Weaver (1952) as typical of the "novaculite-type surface." It is uncertain whether this represents real differences in grain morphology or a difference in quality of replicas. The spongy surface characteristic of chalcedony is absent in our samples. In addition, the spherical objects 0.1 μ in diameter, photographed by Folk and Weaver and interpreted by them as water bubbles, have not been recognized with certainty in our samples. Colored chert beds of Paleozoic age in Japan generally have either well-defined novaculite-type surfaces or spongy textures typical of chalcedony (Kaibara, 1964). Colored chert beds studied by us are described in Table 4.

Subnovaculite

Subnovaculite samples differ in one or more textural or compositional details from novaculite, in addition to their off-white color (Pl. 1), and show considerable variation among themselves. The major differences are as follows:

(1) Subnovaculite is less homogeneous in texture than novaculite (Pl. 7C and 7D). Most samples have layers of UFG and VFG microquartz several millimeters thick, and many have layers of quartz silt up to 2 mm thick.

(2) Subnovaculite contains more impurities than novaculite. In addition to quartz silt, it has more pyrite and hematite particles. Both occur chiefly as grains less than 10 μ long, but locally hematite forms a coating along secondary fracture surfaces.

(3) Some subnovaculite contains lenses and pods of megaquartz that comprise up to 15 percent of the rock.

(4) Subnovaculite has a groundmass in which CG megaquartz is more common than in novaculite.

(5) Fewer clear ghosts of Radiolaria and spicules are visible in subnovaculite. However, spicule canals are locally conspicuous where they have abundant hematite inclusions; a few canals are replaced by platelets of collophane? (pale-yellow mineral that is nearly isotropic, n > C.B.).

The greater inhomogeneity of the texture of quartz in subnovaculite shows in hand specimens as a blotchiness in the shades of white.

Electron micrographs of sample 136 show chiefly equant grains from 0.7 to 2.0 μ in diameter (*see* Table 4).

Tan Chert

Tan chert (Pl. 1) is a common chert type in the lower chert member and locally in the novaculite member where it commonly grades vertically into novaculite. Tan chert is characterized in thin section by:

(1) Minute particles of hematite, limonite, and lesser pyrite and carbonite scattered throughout (Pl. 12C).

(2) Clay in trace amounts or as sparse clots in samples where it was apparently introduced by burrowing animals. The clay is stained light brown by organic matter and possibly by goethite.

(3) Finer grain size than novaculite. It is largely UFG-MG microquartz, much of it very even grained.

(4) Large Radiolaria in some beds. The Radiolaria are clear, circular forms of CG microquartz or chalcedony; no test wall structures are preserved.

(5) Sparse spicules.

Sparse bedding planes are visible in hand specimens, and in some samples clear spots that are Radiolaria can be seen by using a hand lens.

Green Chert

This chert is typically a dull, dark-green rock in which bedding is displayed by lighter or darker colored layers (Pl. 1), rarely by silt laminae. Bright-green beds are present locally and are very colorful. Outlines of Radiolaria can commonly be seen in hand specimens.

Green chert is composed largely of VFG microquartz (Pl. 11A). All samples contain illite, but amounts range from 1 to 33 percent (the arbitrary lower limit of siliceous shale). Radiolaria are conspicuous constituents (5 to 20 percent of the chert because in thin section they appear as clear spots of CG microquartz in the light-brown background of UFG microquartz. Radiolaria show a variety of shapes ranging from circles to very flattened ellipses, depending on the extent of compactional deformation. Spicules are either absent or present in trace amounts. A few samples have layers of silt less than a millimeter thick. Small amounts of pyrite and hematite pseudomorphous after pyrite are present in all samples.

The clay in the green chert is well oriented parallel with bedding and shows conspicuous mass extinction under crossed polarizers.

Two samples of green chert were studied using electron microscopy. The samples (Table 4) are fairly even grained, but differ in modal size. Sample 36 has large, elongate grains up to 20 μ long that are probably clay minerals. A paisley-shaped object, possibly of skeletal origin, was found in sample 37. Chemical analyses of green or greenish chert samples (Table 2) show significant amounts of Al_2O_3, K_2O, and FeO that are present in illite.

Gray and Black Chert

Dark-gray to black chert occurs as (1) thin layers in green-black chert beds, (2) mottles in lumpy, slump-deformed beds, and (3) as rare, coal-black, isolated, boulder-like masses. The dark-colored rocks have considerable variation among samples; some are relatively homogeneous, whereas others are blotchy-textured and mottled (Pl. 11B, 11C, and 11D). Most are composed of UFG and VFG microquartz, but MG microquartz rocks are also present. Thin sections are various shades of brown in transmitted light, but the pigmenting agents for the various colors are not consistent. The blackest cherts are colored by manganese oxide that occurs either as a coating on boundaries of microquartz grains or as minute lacy networks that pervade the microquartz. Some black chert lacks discernible manganese oxide, but is rich in evenly colored dark-brown microquartz whose pigment is either limonite or organic matter.

Gray chert (Pl. 11C) contains disseminated limonite with minor hematite and organic matter. The iron oxides generally replace pyrite and lesser manganiferous calcite. Locally, limonite has bled away from hematite pseudomorphs to strongly color small patches of microquartz.

Radiolaria (Pl. 15) are conspicuous in small amounts in most samples, whereas spicules are less common. Radiolaria occur chiefly as clear circles composed of CG microquartz and chalcedony, but many are flattened. Two beds are exceptionally rich in fossils: one contains 40 percent Radiolaria, and the other contains 60 percent spicules.

Chemical analyses of greenish-gray and greenish-black chert samples (Table 2) show no greater than average MnO_2 content, but greater Al_2O_3, Fe_2O_3, and H_2O than novaculite. A sample from the blackest, boulder-like chert mass has 60 percent MnO_2. X-ray diffraction patterns of black chert show that the MnO_2 is not well crystallized, but patterns have small peaks typical of either pyrolusite or manganite.

Brown Chert

Dark-brown chert is not a common variety, but many beds of other colors have a brownish cast. Beds are chiefly VFG and UFG microquartz pigmented by limonite mixed with manganese oxide and possibly organic matter. Bedding is visible in hand specimen by bands of different shades of brown.

Blue and Red Chert

Beds of vitreous, almost translucent blue to bluish-gray chert are conspicuous in members 4 and 5 in the eastern limb of the Dagger Flat anticlinorium. Samples of the bluest chert have novaculite-type textures, but contain more impurities than novaculite. The only unusual ingredient of the samples is authigenic collophane? (the mineral has high relief, is nearly isotropic, and is pale yellow to light brown). The mineral occurs as pseudomorphs of spicules (apparently canal fillings) and other irregular grains up to 0.07 mm long. A blue-white chert sample that was analyzed chemically (Table 3) has less P_2O_5 than the average chert.

Streaks or lenses of red chert (jasper) occur at only a few localities. Thin plates of translucent hematite produce the color of this chert.

Banded and Mottled Chert

These are strikingly colorful rocks (Pl. 1) that display a variety of textures and patterns. Most formed where chert beds of different color were lithified as a compact bed instead of being separated by shale partings.

A common type of mottled chert is formed by soft-sediment deformation and is diagnostic of the upper part of member 5, but locally occurs in member 3, also. These beds were briefly described in Part 1. They are characterized by uneven bedding planes, many of which have cauliflower-like lumpy surfaces, uneven thickness, mottled colors, and they locally contain limestone and chert clasts and pieces of petrified wood. The beds range from nearly featureless dark-green, brown, or gray chert with scattered patches of chert of other colors, to a maze of varicolored patches cut by white chalcedony and megaquartz veins. Varicolored granule-size chert clasts have distinct boundaries, but large colored particles up to 60 cm long commonly have indistinct boundaries, and it is difficult to determine whether they are clasts or locally stained host chert. Thin sections show the colored fragments have different internal textures and are clearly clasts. Some megaquartz veins are straight and have razor-sharp contacts with host chert, yet elsewhere in the same bed the veins are ptygmatic and have indistinct contacts.

The geometry of the veins and lumpy bedding surfaces indicates deformation occurred when the chert was soft. Brecciated particles produced during slumping subsequently were healed by clear quartz. Examples of slump fractures healed by UFG microquartz also have been found. Throughout the study area the slumped beds locally are greatly enriched in MnO_2, and some are coal black.

An additional type of mottled chert is brecciated nonwhite chert beds that occur in fault zones. These beds generally have been healed by recrystallized or precipitated fracture-filling quartz that is generally white or transparent. Veins of white megaquartz form a network among polygonal fragments of darker colored chert.

Shale

Although shale is an important rock type in the formation, good exposures of it are few. Most samples studied in thin section and by X-ray diffraction are from the road cut 4 miles south of Marathon (Loc. 14) and from East Bourland Mountain (Loc. 19). The samples are mostly from member 5, which yielded beds at least 1 cm thick for thin-section study.

Green and gray-green shale are the most abundant types, followed by light-brown shale. Red shale forms thick beds near the top of member 5 and has a wide areal distribution. White shale that is stained purple by iron oxide along fractures and in scattered spots is prominent in the road cut at Locality 14. The white color may develop by weathering, because white shale is not found elsewhere.

Complete gradations occur between noncherty clay shale to clayey chert, but end-member types are the most predominant. Rocks designated as siliceous shale in the field were found to have at least 33 percent clay.

Shale samples (Pl. 12A, 12B) are chiefly well-oriented illite clay with 1 to 3 percent quartz silt, 1 to 2 percent muscovite flakes, 3 to 4 percent Radiolaria, and variable amounts of hematite specks and organic matter. Quartz silt occurs as subrounded grains scattered randomly in the slide, but occurs as thin laminae (less than 0.04 mm thick) in a few rocks (Pl. 12A). Sedimentation units are defined by layers that differ in silt or radiolarian content: the units range from 0.2 to 8.0 mm thick. A few thin layers are composed entirely of clay except for pigments.

Muscovite flakes less than 0.08 mm long are evenly scattered in the shale. In most rocks they show a highly preferred orientation parallel with bedding, but in a few samples they are largely disoriented. In these beds illite particles are also disoriented and attest to the action of burrowing animals. In most shale samples the illite is well-oriented parallel with bedding and shows pronounced mass extinction under crossed polarizers.

In addition to disoriented mica flakes, evidence of burrowing action was found in a few samples in the form of pods and nests of silt, some with a concentric or swirl orientation, typical of burrow fillings.

Microquartz, where present in the shale, is UFG and VFG, except for Radiolaria. Radiolaria generally occur as clear circles or ellipses of MG to CG microquartz or chalcedony, but in sample X67 (Loc. 19) they are hematite pseudomorphs that preserve excellent details of the capsules. The degree of flattening corresponds in general with the microquartz content of the shale; the less content of microquartz, the greater the compaction (Pl. 14A). In several rocks the capsule walls have been completely collapsed (Pl. 15D).

Spicules are rare. Conodonts and slightly curved plates of probable phosphatic brachiopods occur in red shale, but the shale is devoid of Radiolaria.

Red shale owes its color to hematite that pervades the clay, and brown and gray shale owe their colors to limonite and organic matter. Green shale has trace amounts of hematite and organic matter as the only pigments; the color apparently is imparted by illite (Keller, 1953).

Ten shale samples were studied by X-ray diffraction. Bulk and oriented slides of less than 2 μ particles were X-rayed, and some samples were heated to 550° C and rerun to distinguish chlorite from kaolinite. Data are shown in Table 5. In addition to illite, which is present in all samples, small amounts of kaolinite occur in three samples, montmorillonite in one, magnesium-rich

TABLE 5. MINERALOGY OF SHALE SAMPLES DETERMINED BY X-RAY DIFFRACTION

Q = quartz, I = illite, K = kaolinite, C = magnesium-rich chlorite, M = montmorillonite, ML = mixed layer clay, H = hematite. Minerals listed in order of decreasing abundance.

Sample	Composition
Red shale	Q, I, H
Red shale	Q, I, H. $Fe_2O_3 \cdot S_2O$
Green shale	Q, I
Pale green shale	Q, I, Feldspar
Green siliceous shale	Q, I, K
Brown siliceous shale	Q, I, M
Gray shale	Q, I
Black shale	Q, I, C
Purple-white siliceous shale	Q, I, K, goethite?
White siliceous shale	Q, I, K, ML

chlorite in one, and mixed-layered clay in one sample. Chemical analyses of red and green shale are given in Table 3.

Sandstone

Layers of siltstone and very fine sandstone up to 0.5 mm thick occur in chert beds throughout the formation. Separate beds of sandstone have been found in the lower 10 feet of the lower chert and shale (member 3) at many localities. The beds are light-orange-brown sandstone in separate layers from 1 to 5 inches thick that are laminated (Pl. 13A), and most have fossil burrows on both upper and lower bedding planes (Pl. 13B); a few beds are internally burrowed. No more than six beds have been found at one locality; half of these are megascopically graded (from coarse to very fine sand) and laminated. Chertified coarse-grained sandstone beds a few inches thick are numerous throughout the formation at the western edge of the Payne Hills (Loc. 7).

Thin sections show (Pl. 12D) the beds from member 3 (Loc. 14) are bimodal mixtures of fine-grained quartz and coarse-grained rock fragments, chiefly chert. The composition of individual beds differs only in the relative amounts of quartz and chert. The average composition is: quartz, 60 percent; chert, 35 percent; collophane grains, 2 percent; shale, 1 percent; glauconite, 1 percent; and 1 percent total of phosphate brachiopod(?) shells, K-feldspar, chlorite, zircon, tourmaline, rutile, apatite, and opaque iron oxides. Several beds contain spherical limonite nodules up to

1 cm in diameter; the nodules appear to be alteration products of authigenic iron-carbonate nodules, probably siderite.

The quartz grains are nonspherical, subangular to subrounded particles that are relatively free from mineral and fluid inclusions and have nonundulose extinction. They are cemented by quartz overgrowths. Laminae in the beds are composed of quartz-rich layers alternating with chert-rich layers.

Chert clasts (mostly FG and MG microquartz) are cemented by microquartz of similar grain size, and clasts are discernible only where they differ in color from the clear cement or are surrounded by quartz grains. Most chert grains have ragged boundaries that are due either to a soft condition at the time of deposition or uneven replacement of framework grains by cement. The grains are of coarse-sand size, elongate and are well aligned parallel with bedding.

Collophane grains have uneven borders; they appear to have either grown in place by replacement of other clasts or enlarged their size by authigenic growth.

Although the overall sorting of the sandstone beds is poor, the sorting of the separate modes is moderate.

Goldstein (1959, p. 147) also described sandstone samples collected from Locality 14.

Conglomerate

Granule conglomerate beds are conspicuous throughout the area studied, although only two or three are found at any one locality. They are locally present, except in the novaculite members, but are most abundant in member 5. Areally they are thickest and most abundant along the north, east, and western margins of the Marathon Basin.

Beds range from a millimeter to 4 feet thick, but most are less than 3 inches thick. Beds are varicolored because of the different colors of chert clasts, and most are faintly graded. Conglomerate beds are of uneven thickness along strike, and commonly vary by a factor of two over a distance of 20 feet. At Rock House Gap (Loc. 10), a bed 22 inches thick becomes only 2 inches thick 100 feet along strike. Grains are cemented by FG to CG microquartz (minor chalcedony) to yield resistant beds.

Like sandstone, the conglomerate is bimodal and has an over-all poor sorting. The coarse mode is elongate chert and shale clasts 1.0 to 4.0 mm long and a minor quartz mode of fine sand (0.2 mm). The clasts (Pl. 14C) have an average composition of: chert, 60 percent; shale, 15 percent; quartz, 15 percent; collophane, 7 percent; and the remainder of glauconite, siltstone, and brachiopod(?) valves. Trace amounts of K-feldspar and heavy minerals are present. Two samples contain one grain of altered volcanic rock, one sample of which also has a few biotite flakes. Several samples have conodonts.

The many varieties of chert clasts present include all chert types of the Caballos Novaculite. Organic-rich chert with small spicules are abundant, and some may be from the underlying Maravillas Formation. Most clasts have been unevenly replaced by microquartz cement to produce ragged boundaries; hence, the original roundness of grains is not clearly shown. Most grains appear to have been subrounded to rounded.

Calcarenite

Calcarenite beds range from 3 to 18 inches thick and are light gray weathering to light brown. The thinner beds are graded and have laminae (Pl. 14B) in the top half of the bed. A few beds are free of secondary chert, but several have been largely replaced by CG microquartz.

Pelmatozoan fragments (mostly crinoids) and micrite rock fragments in a ratio of 2:1 to 3:1 comprise about 70 to 90 percent of the calcarenite samples. The remainder of the clasts are composed of chert, calcite, and phosphatic brachiopods, bryozoa, algae, ostracods, mollusks(?), trilobites(?), glauconite, quartz, K-feldspar, and collophane. Calcite spar and microspar and, locally, microquartz, are the cements. The micrite clasts, some of which have embedded fossils, are well rounded, but most fossil fragments show only incipient rounding. Authigenic dolomite or ankerite occur in most beds in trace amounts.

Comparison with Previous Studies

Thin sections of Caballos samples have been described briefly by several previous investigators. Baker and Bowman (1917, p. 93-101), Goldstein (1959), and Park and Croneis (1969) made comparisons of samples from the Arkansas and Caballos formations, whereas King (1937, p. 51-52), Bennett (1959, p. 71), and Thomson (1964, p. 12) described only the Caballos. The authors use different terms to describe their chert samples, so that the over-all descriptions yield slightly different mental pictures.

Baker and Bowman (1917, p. 99) describe novaculite samples ranging from ". . . fully as fine and uniformly grained as any of the Arkansas commercial novaculites . . . ," and other samples having a very uneven texture. Park and Croneis (1969, p. 100) describe the novaculite matrix as cryptocrystalline, King (1937, p. 51) as uniformly fine-grained siliceous material, and Goldstein (1959, p. 147) as uniform and composed largely of cryptocrystalline to microcrystalline quartz with subordinate amounts of chalcedony. The scattered quartz grains were mentioned by all the above authors, and the particles we call specks were mentioned by King, Bennett, and Park. Most investigators either overlooked the spicules or Radiolaria in novaculite or studied samples where they are not discernible. Exceptions are Baker and Bowman (p. 100-101), who described them to range from well-preserved tests to "knots" of coarse recrystallized quartz. Park and Croneis (1969, p. 101) describe 13 of 24 thin sections of thin-bedded chert from both the Arkansas and Caballos formations to have oriented cryptocrystalline silica, such that slides show a pronounced mass-extinction effect. In slides studied by us, illite is the only mineral that shows preferred orientation; microquartz is oriented randomly. Park and Croneis (p. 98) also report that slides of Caballos chert show evidence of slump structures and possible cross-stratification. Although cross-stratification is not evident in samples studied by us, a few rare examples of laminae truncated at low angles, presumably by erosion, are present.

Goldstein (1959, p. 147) noted that the more impure chert beds have a finer particle size than novaculite and that finely divided sapropelic organic matter is apparently effective in limiting the particle size of microquartz. He mentions also the presence of spore exines in some chert beds. Objects the size of those figured by Goldstein (1959, Fig. 15) as spores are relict radiolarian capsules, according to our studies.

Silt and sand layers in the Caballos were described by King (p. 52), Goldstein (p. 147, Fig. 14), and Park (1961, p. 23).

Areal and Vertical Stratigraphic Differences

Areal and vertical stratigraphic differences in the formation are few, except for the megastratigraphic details described in Part I of this report. However, the following were noted:

(1) The lower chert member contains considerable terrigenous material in the form of laminae of coarse silt and very fine sand-size detritus. The laminae are generally less than a millimeter thick and are composed predominantly of quartz with trace amounts

of muscovite, heavy minerals, glauconite, and clay. The volume of terrigenous grains in the member nowhere exceeds 15 percent of individual beds.

(2) In the lower chert member, contorted laminae of terrigenous silt and sand and local clay-rich layers are particularly prominent in the northern part of the western limb of the Dagger Flat anticlinorium. The contortions developed during a soft-sediment stage of deformation by either slumping, burrowing, or escape of gas.

(3) The greatest percentage of terrigenous and detrital chert grains in the formation occurs (a) in the Payne Hills, and (b) in the northeasternmost exposures. Very fine sand, silt, and clay occur throughout the formation in the Payne Hills; in fact, all chert beds contain at least 1 percent terrigenous grains. Many granule conglomerate beds and slump-contorted chert beds that locally bear pebbles and cobbles are present in the northeasternmost exposures.

(4) Although most samples show the effects of structural deformation, generally the strongest recrystallization and veining by chalcedony and megaquartz is around the southern and western margin of the Marathon Basin.

(5) Typical novaculite (milk-white chert) is best developed along the eastern edge of the Marathon anticlinorium.

(6) Samples from the Solitario Uplift are similar in general to those in the Marathon Basin. However, specks in novaculite from the Solitario have sharper boundaries with the host chert, and their relict radiolarian origin is more convincing.

SEDIMENTARY AND DIAGENETIC FEATURES

Within the Caballos are a number of features of either primary or secondary origin that have not been described, but which are important in interpreting the origin of the rocks. They range from features of microscopic size to outcrop details and include the following: pseudo ripple marks, turban concretions, pits and mounds, trace fossils, Liesegang bands, pillow structures, geopetal fabric, knobby bedding surfaces, slumped and MnO_2-stained beds, and several miscellaneous features.

Pseudo Ripple Marks

Bedding planes marked by rhythmic undulations (Pl. 16A) that resemble water-formed ripple marks are locally prominent in the lower novaculite member where they were noticed by Baker and Bowman (1917, p. 95), King (1937, p. 49) and Bennett (1959, p. 52-55). On any single surface, ripples are remarkably uniform in geometry (length, height), have straight ridges that locally bifurcate, and generally have symmetrical crests. Most ripples are exposed on the soles (undersides) of beds; exposed matching bedding surfaces are rare. Previous investigators interpreted them as ripples of sedimentary origin [although Baker (*in* Davis, 1918, p. 335) suggested they may be folds], but for several reasons we believe they are diagenetic features. The following observations pertain to the ripples:

(1) Most beds clearly show that the undulose rippled surfaces presently exposed are stylolite surfaces. The amplitude of stylolite sutures is microscopic, however, and insignificant in comparison to the ripple amplitudes.

(2) The crests of some ripples are situated along veins of vitreous microquartz that are zones of recrystallized novaculite.

(3) Some ripples have a form unlike sedimentary ripples in that troughs are too flat or wide compared to the crests, or too deep (Fig. 12) for the wavelength. Wavelengths for eight beds range from 0.3 to 5.0 inches and average 2.9 inches (7.3 cm). In addition, ripple crests are generally straighter than water-formed ripples.

(4) In a succession of three to five beds with ripples, the orientations on the beds are generally coincident. This is greater uniformity than shown by most sedimentary ripple marks.

(5) The orientation of 23 sets of ripples from localities along adjacent limbs of the Dagger Flat and Marathon anticlinoria have a bimodal distribution (Fig. 13). Modes at N. 60° E. and N. 60° W. form a 60° to 120° complement. The major modes fit the orientation of a shear set for a principal stress oriented east-west. This direction does not coincide with the N. 50° W. orientation of the principal stress that produced the folds, thrusts, and strike-slip faults in the basin. The ripples, with two exceptions, also are athwart the direction of prominent joint sets in novaculite outcrops in which the ripples occur.

(6) The ripples are not microfolds. Bedding planes adjacent to rippled surfaces are undeformed and show that the ripples are not small compressional wrinkles.

From the evidence available, it is concluded that the ripples are secondary features that formed by an unusual style of stylolite

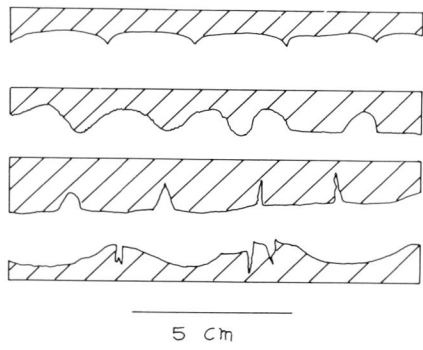

Figure 12. Cross sections of pseudo ripple marks in novaculite, from field sketches and photographs. The bottom pseudo ripple is on an upper bedding plane, whereas the others are sole marks.

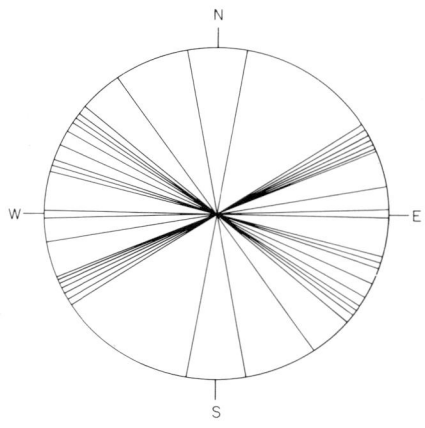

Figure 13. Orientation of pseudo ripple marks in novaculite.

development. The possibility that the stylolites merely modify wave- or current-formed sedimentary ripples is unlikely in view of the strong 60° to 120° complementary orientation that suggests structural control. In addition, the several beds with ripple crests that coincide with the position of veins of recrystallized microquartz indicate a diagenetic control on the position of ripple crests. It is inferred that the stylolites formed under lithostatic load when the chert was subjected to an east-west stress prior to the major orogenic episode of crustal shortening.

Turban Concretions

Squat, subspherical objects shaped much like coiled turbans (Pl. 17B) have been found in the lower novaculite at Localities 11 and 18 at positions of 15 and 60 feet above the base of the member. They are best developed at Locality 18, where at least six concretions are exposed along two bedding planes. The turban-shaped objects are 12 to 16 inches in diameter, 6 to 10 inches high, and elongate in the plane of bedding. Novaculite chips have spalled off the objects along uncurved fractures to produce nicely rounded forms. Several concretions have central depressions up to 2 inches deep, the walls of which are circular stylolite seams. The core of one exposed concretion is an area 2 inches in diameter in which angular pieces of novaculite are cemented by drusy megaquartz.

The origin of the objects is not clear. Apparently a solution pocket developed in novaculite, and when lithostatic or tectonic stress, or both, was applied, the novaculite strained in a unique pattern controlled by the cavity (or cemented cavity). The vertically oriented concentric stylolite probably originated following or simultaneous with development of the central cavity.

Pits and Mounds

Circular pits and mounds of a wide-size range occur throughout the formation, but are most abundant in novaculite, the lower chert member, and on white chert and subnovaculite of other members. Small pits and mounds less than 1 inch in diameter occur in all the above units, but the large ones are found only in novaculite.

The most common large form is a circular pit on the top bedding surface, and most pits are surrounded by a raised rim. At present, pits extend as deep as 2 inches below the rim. Plate 18 shows the range in shape of the largest structures found. Casts of pits and mounds on the sole of beds are about equally common as the structures on the tops of beds. As far as can be determined, shale was the bed immediately adjacent to the chert surfaces that have the structures. The profusely covered sole of Plate 18A shows the impression of pits with concentric ornamentation. The novaculite bed immediately below it has pits that match up with only the central part of the circular rings. This shows that the novaculite sole mirrors the ornamentation of a thin mud layer (shale parting) that has spalled off during weathering.

The origin of the pits like those illustrated is uncertain: they formed either by escaping gas or by burrowing organisms. Some mounds (Pl. 18B) resemble those photographed from the floor of the Atlantic and attributed to organisms. However, those pictured in Plate 18A are believed to be gas-formed because they are perfectly symmetrical, have annular ornamentation, and occur on bedding surfaces that are devoid of trails or other traces of bottom life. Beds were examined to see whether the path of moving gas could be traced by disturbed laminae, but the largest structures are not amenable to such study, and sawed slabs of smaller pits are devoid of internal bedding. Thin sections of chert from the lower chert member show disrupted laminae that could have formed by either gas or organisms.

Trace Fossils

Spoor left by benthonic animals are present in the form of burrows perpendicular to and parallel with bedding, and the rare fecal pellets that have been recognized in novaculite. Burrows have been found in the bedded chert of members 1, 3, and 5, and in sandstone and shale, but not in the novaculite members with certainty. The most widespread trace fossil is the scar of a vertical burrow from 2 to 4 mm wide seen on a bedding plane as a circular dimple that lacks a raised lip; raised tubes are rare. Short segments of burrow trails parallel with bedding (Pl. 13B) are much less common. Trails occur on the surface of chert beds separated by shale partings; this suggests the burrowing animals sought food in the mud layers preferably to the siliceous layers.

Evidence of disturbance by animals is present in thin sections of some chert (Pl. 14D) and shale beds in the form of disrupted silt layers, disoriented clay minerals, and disoriented spicules. Beds with conclusive evidence of burrowing comprise an estimated 5 to 10 percent of the section, and the burrow density is low in any one bed. Rather intensely burrowed beds occur in a succession of a few tens of feet of section at most outcrops.

Knobby Bedding Surfaces

Knobby bedding surfaces are locally prominent in colored chert beds of members 3 and 5, and are developed to a remarkable degree along the western margin of the Marathon Basin. At Locality 8 every bedding surface of colored chert (the novaculite members are absent) in the lower third of the formation is rough with knobs (Pl. 16B).

In members 3 and 5, the knobs have less relief than at Locality 8, but their form is similar. The knobs are in chert beds that are separated by thin shale beds and partings. Knobs range in width from 5 mm to 10 cm (one occurrence of 15 cm is known) and in relief from 1 to 40 mm. The larger knobs are broad, gently rounded domes, whereas the smaller knobs are more peaked. Large domes occur only on upper bedding surfaces, but counterparts of small knobs on overlying chert beds (separated by shale) resemble upper surfaces to a considerable degree. Generally, the relief is less on knobby soles than upper surfaces.

The smooth surface of many knobs, which apparently was a bedding plane, protrudes through adjacent layers in diapir fashion. In fact, the knobs formed by the upward bulging of local masses of what is now chert. Some bulges formed by deformation of the entire upper part of a proto-chert bed, whereas others began inside the bed and punched through the upper laminae. The geometry indicates the knobs formed when covered by at least the thickness of a shale layer, and probably by several beds of proto-chert.

At Locality 8 the knobs are pale red and are encased with fuzzy boundaries in light-green chert to produce a colorful mottled chert (Pl. 1).

Knobs of similar character to those in the Caballos have been described from chert in the Schoonover Formation in Nevada by Kirchmayer (1959) and Fagin (1963). Both authors infer the knobs formed by diagenetic flowage of laminae of density different from host chert, but the cause of the density difference is speculative. The Caballos examples shed no new light on the problem.

Liesegang Bands

Liesegang bands of pale-red to orange iron-oxide stains are locally present in the novaculite and chert beds of the formation. Some bands follow joints, whereas others do not. The time of band development spanned the time of jointing of the rocks.

Geopetal Fabric

Several beds of novaculite and subnovaculite have spherical or elliptical vitreous "eyes" of megaquartz, from 1 to 3 mm in diameter, that are floored by organic-stained microquartz. After the formation of spherical cavities, insoluble material accumulated along their lower surface, and the cavities were later filled by megaquartz. The level floor of the geopetal "eyes" is parallel with bedding and attests to a prefolding stage of formation.

Slumped and MnO_2-Stained Beds

These beds were described briefly as characteristic of the upper chert and shale member, but deserve additional mention because of their widespread occurrence, unusual texture, and significance.

In most outcrops, the one or two mottled beds in the section stand out from adjacent strata because they are slightly thicker than adjacent chert beds, they resist weathering more than adjacent beds, and they are mottled rather than even-colored. Most beds, although of uneven thickness, range from 6 to 24 inches thick. Locally, however, greater thickness variations exist, and beds attain thicknesses of 15 feet. The thickest beds, and some thinner ones, are deeply impregnated by MnO_2 and look like blocks of coal embedded in the outcrop.

The largest and several smaller black chert "blocks" crop out in the valley of Little Woods Hollow syncline. The largest occurs on the west limb of the syncline (Loc. 19) 22 feet above the upper novaculite (here only 8 feet thick), and is a resistant block 34 feet along strike and 15 feet thick (separated by a joint into two chunks). No distinct bedding planes are discernible, but what appear to be faint ones are undulose. MnO_2 pervades the host rock unevenly: some areas are coal black, others have patchy remnants of unaltered white chert. The rock is compact, except where soft, clinkery patches of MnO_2 have weathered out. The block is laterally equivalent to light-colored chert and shale along strike that are only slightly stained by MnO_2; it is uncertain whether the block is a submarine slide block or altered host rock. A sample from this block is 60 percent MnO_2.

About a quarter mile north on the east side of the valley is another black-stained bed 58 feet above the lower novaculite member. The bed ranges from 6 to 24 inches thick, has a lumpy top surface, and rests on slump-contorted chert beds in a zone 1 foot thick. The stained bed is clearly slumped and not exotic. Above a 2-foot covered zone is a second mottled bed, here unstained by MnO_2, that is 12 inches thick, blotchily colored by hematite, and locally internally brecciated and invaded by white chalcedony that cements the rock. Several thinner mottled beds occur higher in the section. The top surface of the upper novaculite member here is deformed into an undulose surface with 6 inches relief, and green chert fills small fractures that formed during slumping.

Another conspicuous resistant bed locally crops out as a large block 100 yards northwest of the northwest edge of the Combs Ranch klippe. This mottled bed can be traced for considerable distance along strike; it is generally 16 to 24 inches thick, but is 9 feet thick where the "blocks" are exposed. The bed is 120 feet above the upper novaculite (4 feet thick here). The block is devoid of bedding, is colored mottled gray, green, red, and white, and is locally veined by chalcedony. MnO_2 stains some joint surfaces and local patches of the block. The bed occurs between thin green chert and shale beds of this upper member.

The mottled beds described above are interpreted as beds deformed uring submarine slumping and not during a tectonic event because of the following evidence:

(1) The beds have wide stratigraphic continuity.
(2) Beds above and below are normal in appearance and are undeformed.
(3) Beds with lumpy surfaces and uneven thickness are common features of soft-sediment deformation.

(4) The beds commonly contain scattered clasts of limestone and chert that were deposited during a catastrophic event.

The brecciation shown by the beds does not extend into adjacent strata. The brecciation formed when slightly consolidated chert beds fractured during slumping, and the breccia pieces were subsequently recemented by chalcedony (Pl. 1).

Iron-Oxide Nodules

An additional occurrence of concentrations of iron oxide is in nodule-like bodies in siliceous shale and green chert. The objects are disc-shaped or irregular in form, range from 1 to 20 mm in diameter, and have a rind of relatively compact iron oxide enclosing a vuggy or boxwork interior of similar composition (Pl. 17A). Most nodules are filled with a mixture of hematite and goethite. One nodule contains euhedral crystals of martite (hematite pseudomorphous after pyrite) up to 2 mm in diameter, which suggests that all nodules originally may have been pyrite. Low concentrations of hematite in siliceous shale impart a purple color that suggests the presence of MnO_2; however, a spectrographic analysis of one purple sample shows less than 0.01 percent MnO_2, but 3.0 percent total Fe as Fe_2O_3. The pods occur chiefly in the upper chert and shale member.

Miscellaneous

Stylolites have been described previously because they are common between beds of novaculite; they are rare in colored chert. Almost all stylolites are parallel with bedding, but a few have been found that make low angles with bedding, and concentric stylolites occur in the turban concretions. Measurements of the amplitudes of stylolites in one stylolite-rich area indicate a 15 percent loss in stratigraphic thickness by solution in this sequence.

Two additional bedding-plane irregularities deserve mention because they are common. Where large areas of bedding surfaces of novaculite or subnovaculite are exposed in oblique light, broad undulations of very low relief are visible. These undulations resemble current ripple marks in pattern, but are much too broad (wave lengths of 12 to 16 inches) compared to depth (1/16 inch). On the same beds and on additional beds are wrinkle-like linear irregularities which commonly occur in patterns — intersecting sets or incomplete radial patterns. The ridges have scarps up to 1/4 inch, but most are less. Baker and Bowman (1917, p. 95) referred to "hummocky" bedding surfaces in the novaculite and probably saw similar features.

The latter ridges are thought to be scars of small fractures with minute dislocations that healed soon after formation. The ripples are also probably postdepositional in origin. Although both features bear no obvious relation to major tectonic patterns, they are best developed in areas of greatest deformation around the margin of the basin.

PETROGENESIS

Review of Origins of Other Bedded Chert Formations

The problems of the origin of the Caballos are similar to many that relate to the origin of bedded chert in general. Pettijohn (1957, p. 435-444) reviewed the general problem of the origin of bedded chert formations with emphasis on those in North America. Grunau (1965) recently summarized the worldwide distribution and occurrence of Radiolaria-bearing bedded cherts with emphasis on those in the Tethys geosyncline. He also summarized the problems of origin and the prevailing views of European workers.

Although bedded chert formations have considerable variation in detail, as a group (omitting the Precambrian iron formations) they have a similar tectonic setting; they are deposited in linear geosynclines and precede deposition of the main filling phase (flysch deposition). Europeans have generally considered bedded chert as deep-water deposits, whereas, until recently, most North Americans have favored a shallow-water interpretation. Recognizable remains of siliceous organisms comprise only a few percent of most formations; some workers interpret them as incidental constituents, whereas others maintain they are relicts of what was originally entirely siliceous ooze. Submarine volcanic rocks are associated with many chert formations, and volcanism commonly is considered to be the source of dissolved silica in sea water that is subsequently precipitated inorganically or extracted organically. However, some chert formations are not associated with volcanic rocks, and have no evidence of pyroclastic contributions. This has led to the conclusion that these formations are the products of accumulation of siliceous ooze in normal, deep-marine basins that received almost no terrigenous sediment.

Siever (1962) emphasized that silica will not spontaneously precipitate from normal sea water; and that sea water would have to be reduced in volume over 100 times before saturation with respect to amorphous silica is reached.

If the concentration of silica in sea water in the past has not differed greatly from its present concentration (1 to 4 ppm), inorganic precipitation could occur only under unusual local conditions. Such an occurrence is postulated by Bailey and others (1964), for the origin of the bedded chert of the Franciscan rocks in California, whereby silica released by submarine volcanic rocks at oceanic depths is dissolved at high temperature and pressure and subsequently precipitated during ascent by convection. However, the absence of evidence of volcanic activity requires a different explanation for formations like the Caballos.

The chief objection offered to the hypothesis of chert formed as thick deposits of siliceous organic ooze not associated with volcanic rocks is the uncertain source of silica. Most workers have argued that siliceous organisms could not accumulate fast enough or in sufficient quantity to form thick, bedded chert formations if the only source of silica is that present in normal sea water. However, Calvert's (1966) study of biogenic silica deposits in the Gulf of California shows this objection is no longer tenable.

Although origin of the Precambrian banded iron formations has some problems in common with the origin of the geosynclinal chert units under discussion, the two types of deposits differ considerably in mineralogy and geologic setting. Evidence is steadily being marshalled to support a lacustrine origin for the banded iron formations (Hough, 1958; Govett, 1966; Eugster, 1967), whereas the Caballos and related chert formations are clearly open-marine deposits.

Review of Hypotheses of Origin of the Arkansas and Caballos Novaculites

The Arkansas Novaculite of the Ouachita Mountains correlates lithologically and, in part, temporally with the Caballos Novaculite, and both were deposited in the greater Ouachita geosyncline. Although this study is confined to the Caballos Formation, it is pertinent to review the hypotheses of origin proposed for the Arkansas Novaculite in addition to the Caballos because of their similarity, and because the Arkansas Novaculite has been studied more extensively than the Caballos.

Detailed reviews of possible origins of the Arkansas Novaculite are given by Griswold (1892, p. 169-194), Miser and Purdue (1929, p. 49-59), Park (1961), and Park and Croneis (1969). The views of authors are summarized in chronological order in Table 6.

TABLE 6. ORIGINS PROPOSED FOR THE ARKANSAS NOVACULITE

Owen (1860, p. 23-24): sandstone altered by heated alkaline siliceous water.
Branner (*in* Comstock, 1888, p. 49): novaculite is selectively metamorphosed bedded chert.
Comstock (1888, p. 95, 129): alteration of quartz (sandstone implied) in place by hot water.
Griswold (1892): very fine clastic quartz.
Rutley (1894): siliceous replacement of dolomite or dolomitic limestone.
Hinde (*in* Rutley, 1894, p. 391-392): probably organic silica.
Derby (1898): replacement of limestone by silica.
Weed (1902, p. 84): chemical precipitate in deep sea.
Van Hise (1904, p. 849, 853): organic precipitate, now recrystallized.
Honess (1923, p. 138-139): in part silicified and devitrified volcanic ash, but mostly chemical precipitate without the aid of organisms.
Miser and Purdue (1929, p. 57): chemically precipitated.
Henbest (1936, p. 78): chemically precipitated.
Hendricks and others (1937, p. 11): initially siliceous deposit derived in part from siliceous organisms.
Harlton (1953, p. 792): replacement of limestone by ground water upon rupture of rocks during tectonism.
Goldstein and Hendricks (1953): submarine alteration of volcanic ash to produce opaline or "isotropic" silica; minor contribution from siliceous organisms; diagenetic recrystallization to chalcedony and cryptocrystalline silica.
Goldstein (1959, p. 149): same as Goldstein and Hendricks.
Park (1961): recrystallized amorphous silica deposited by organisms.
Park and Croneis (1969): same as Park (1961).

Most authors cited proposed origins for the novaculite, but did not comment on the origin of the colored, bedded chert.

The hypotheses of novaculite origin may be grouped into the following (omitting Comstock's hypothesis):

(a) replacement of limestone by silica,
(b) diagenetic or metamorphic alteration of clastic quartz,
(c) inorganic precipitation of silica,
(d) organic precipitation as skeletal particles, subsequently recrystallized or altered,
(e) sea-floor alteration of volcanic ash.

Although the Caballos Formation has not been studied by as many workers as has the Arkansas Novaculite, interpretations of its origin also differ considerably.

Baker and Bowman (1917, p. 100-101) studied thin sections of the formation in conjunction with their reconnaissance study. They suggested that novaculite chert was

... originally made up of amorphous silica which has reached its present form by a process of crystallization analogous to that of the devitrification of extrusive igneous rocks. The finer and more-uniform-grained commercial varieties were either originally a pure siliceous ooze or else a radiolarian chert in which all evidence of Radiolaria has been removed by subsequent metamorphism (recrystallization) .... Since the Caballos is entirely a chert formation and pure siliceous oozes containing radiolarians are known to form only in the deep sea, it is most probable that it is a true pelagic formation.

They noted that "the bedded colored chert has a nearly similar texture as novaculite, but contains more and larger Radiolaria than novaculite." The reference to "commercial varieties" is borrowed from terminology applied to whetstone-grade material of the Arkansas Novaculite.

Baker (*in* Davis, 1918, p. 335-336) is later cited to have supported Branner's (*in* Comstock, 1888, p. 49) concept, and proposed that the novaculite members of the Caballos formed by metamorphic recrystallization of the colored, bedded chert members.

King (1937, p. 54-55) concluded that ripple marks in the novaculite indicated shallow-water oscillation that moved particulate matter (not gelatinous or colloidal material), that a small part of the rock is detrital and organic (quartz and Radiolaria), and that colored chert beds recorded more fluctuations in conditions of sedimentation in a region of quiet water than novaculite. He (p. 55) concluded that because of meager evidence:

... No definite interpretations can be made as to the origin of the novaculite. A relation between novaculite deposition and the secretion of silica by such organisms as Radiolaria, with volcanic activity, is possible but remains to be proved. To the writer the northwestward thinning of the novaculite members, their relations to the banded chert members, and the ripplemarked bedding surfaces suggest that the novaculite of the Caballos formation may have been laid down as a fine clastic sediment, rather than as a precipitate.

Aberdeen (1940) studied Radiolaria in the upper chert member and interpreted the forms to be pelagic, but of shallow-water habitat because of capsule morphology. She notes, however, that the assemblage contains some genera known today only from great depths.

Bennett (1959, p. 71-76) believed the formation probably contained some orthochemically precipitated chert, but thought the presence of fecal pellets and mound structure in novaculite resemble those of modern carbonate mud, and suggested the replacement of limestone as the chief origin.

Goldstein (1959, p. 149), from petrographic study of both the Arkansas and Caballos formations, reiterated his views expressed with Hendricks (Goldstein and Hendricks, 1953) that:

... the source of silica was extensive volcanic ash falls during a time of relative peneplanation when little detrital sediment was being supplied to the Ouachita geosyncline. Submarine weathering, prolonged over long periods of time, removed readily soluble elements and converted the ash to nearly pure opaline silica. Radiolarians and siliceous sponges thrived in the high-silica environment, their remains are preserved in the siliceous sediments. Normal marine conditions prevailed during most of the time these sediments were being deposited, but there were periods in which the bottom waters were stagnant, and high-silica sapropelic sediments containing spore exines were laid down. The only important role that can be assigned to purely epigenetic processes is that of redistribution of the silica already present in these fascinating rocks.

In the report on the Ouachita System (Flawn and others, 1961) both Flawn and King accept Goldstein's ideas that the silica of the chert was derived from volcanic ash falls. Flawn states (p. 52) that the Marathon area during deposition of the chert and shale sequence of the early Paleozoic was one of low or restricted sediment supply with volcanic activity furnishing the silica in the form of ash. King noted (*in* Flawn and others, 1961, p. 181) the similarity of the environment of deposition of early and middle Paleozoic rocks of the Ouachita geosyncline to the leptogeosynclinal deposits of parts of the Alpine geosyncline, but concluded the Ouachita environment was probably shallow rather than deep as proposed for the Alpine trough.

Park (1961) studied the Arkansas and Caballos formations using thin sections and electron micrographs and discussed sedimentological problems concerning the origin of the formations in detail. His conclusions as stated originally (1961, p. 72-73) and recently (Park and Croneis, 1969) are summarized below:

(1) The silica of the formations is primary and precipitated by an organic agent. The original material was probably amorphous silica in the 0.1 to 5.0 μ range.

(2) The accumulation took place in a marine environment that was, in part, reducing. The depositional site received little terrigenous material, and the depositional surface was at times well below wave base.

(3) The fauna of the formations is predominantly pelagic, suggesting that bottom conditions were not suited to benthonic life.

(4) The alteration of precipitated silica to a mosaic of small quartz crystals was accomplished by lithostatic rather than metamorphic pressure. During diagenesis, lithostatic load seldom exceeded hydrostatic load.

(5) The redistribution of silica was probably largely accomplished by connate water with little pressure solution.

(6) The formations represent a time span of at least one geological period (Devonian); thus, the rocks were deposited at an extremely slow rate of deposition.

(7) The concentration of silica in the formations can be explained by a normal rate of accumulation of silica undiluted by other material. There has probably been some contribution to the silica from volcanic sources, but its importance could not be estimated from data obtained in the study.

Fan (1964), in an abstract, proposed dividing the Caballos into three formations of unrelated origin. He states (p. 56):

... the basal white novaculite originated as diatomaceous earth; the middle shale unit as a fluvial deposit of fossil soil and weathered volcanic product; the upper cherty and siliceous shale unit as a submarine alteration product of a sodium-rich, fine-grained volcanic rock.

Summary of Facts Pertinent to the Origin of the Caballos Novaculite

Before a discussion of the origin proposed here, several facts pertinent to an interpretation of the Caballos are summarized below:

(1) Textures and structures and facies distribution of the formation are those of sedimentary rocks, not metamorphic rocks.

(2) Siliciclastic terrigenous material (quartz, clay, heavy minerals) in unaltered and unreplaced comprise only 8 to 10 percent of the formation.

(3) Calcarenite grains in unaltered and unreplaced form comprise less than 1 percent of the formation; but even where replaced by microquartz, the calcarenite texture is recognizable. The absence of calcarenite texture in chert indicates the chert beds did not form by the replacement of calcarenite limestone; this does not exclude the possibility of replacement of carbonate mud, however.

(4) Ninety percent of the formation is chert, 40 percent is novaculite, and 50 percent is colored, bedded chert. Only about 12 percent of the novaculite is demonstrably of organic origin;

only about 6 percent of the bedded chert is demonstrably of organic origin.

(5) Radiolaria and spicules, probably of sponges, are the only identifiable silicic skeletal remains in novaculite, except for one sample with relict nannoplankton of uncertain affinity. Although most novaculite beds have less than 12 percent recognizable skeletal grains, these beds are identical in physical properties with the few beds that have from 70 to 80 percent recognizable skeletal grains.

(6) There is no direct evidence of volcanic contributions in the form of volcanic ash or beds with relict shards; nor is there indirect evidence in the form of abundant accessory volcanic minerals, such as euhedral zircon or apatite.

General Interpretation of Present Article

Major interpretations made here are itemized below. Additional special problems are discussed following a comparison of our conclusions with those of previous authors.

(1) Sediments were deposited in a linear marine basin throughout Silurian, Devonian, and perhaps part of Mississippian time.

(2) Silica in the formation was derived from accumulations of opaline skeletal grains. Radiolaria and sponges apparently contributed most of the silica, but other organisms difficult to preserve, such as silicoflagellates, and other nannoplankton with opaline particles may have made contributions. Silicoflagellates are not known earlier than Mesozoic time, however.

(3) The opaline particles were altered diagenetically during burial and compaction; most skeletal grains lost their structure.

(4) Terrigenous clay settled from suspension. Almost no clay was available during deposition of proto-novaculite. The cause of the cyclic alternation of chert and clay is unknown; it could be due to tectonism, climate, changes in ocean circulation, or other causes, but it is interpreted to be a depositional and not a diagenetic feature. There is no evidence to suggest the clay layers formed by the expulsion of clay from original clay-silica gel mixtures, as proposed by Davis (1918, p. 393-402) for the Franciscan Formation.

(5) Beds of calcarenite, granule conglomerate, sandstone, and siltstone laminae were deposited by turbidity currents. Some silt laminae may be storm suspension deposits. Scattered silt may be wind-blown grains, but more likely are grains of original silt layers dispersed and mixed with the opaline biogenic remains by burrow-

ing animals or by gas escaping through beds. Most clastic layers might be turbidites because they are graded, and are intercalated with possible deep-water pelagic sediments (*see* below).

(6) The colors of the chert beds in the formation are due to a combination of detrital or chemical impurities present in variable amounts (illite, organic matter, pyrite, hematite, limonite, manganocalcite, collophane, and MnO_2). Novaculite is colorless because it lacks such impurities, whereas it probably owes its milkiness to the scattering of light by abundant minute water-bubble inclusions (Folk, 1965, p. 80) and myriads of reflecting faces of microquartz grains.

Although spherical inclusions, 0.1 μ in diameter, such as described and illustrated by Folk and Weaver (1952), are not visible in our electron photomicrographs, larger ramose inclusions with pale-brown color and negative relief are visible in most thin sections of novaculite. The inclusions are presumably water, although this cannot be proven. The brown "specks" of UFG and VFG microquartz in novaculite probably are richer in water inclusions than the remainder of the rock. Chemical analyses of novaculite samples (Table 3) show the H_2O content ranges from 0.02 to 0.14 percent. Float pieces of smoky-gray chert have been found with patinas of milky chert from 1 to 3 mm wide and indicate that the milkiness can form by diagenetic (in this case by weathering) processes.

Comparison with Previous Interpretations

The major thesis of Baker and Bowman that the formation is a pelagic deposit and that novaculite was originally a siliceous ooze is accepted. Baker's later idea that novaculite formed from colored chert by metamorphic recrystallization is rejected because: (a) the novaculite and chert and shale members are clearly stratigraphic units rather than metamorphic facies, and (b) novaculite had its present appearance before the development of joints and before the recrystallization that followed tectonic brecciation. Baker apparently thought recrystallization was a metamorphic process, but this recrystallization could better be attributed to an early diagenetic stage.

King's early suggestions about origin were made prior to modern theories on geosynclinal deposition and suffer from this lack of information. His suggestion that proto-novaculite was a clastic (and presumably terrigenous) deposit is not supported by the evidence presented here; but his interpretation that protonovaculite was particulate grains instead of a precipitate is sound. The interpretation that novaculite, colored chert, and shale were deposited successively farther from shore reflects the old idea of sand-shale-carbonate facies tracts. The interbedded character of the lithic units refutes such an interpretation. King (*in* Flawn and others, 1961) later changed his views to support Goldstein's hypothesis of the submarine alteration of volcanic ash.

Goldstein's (1959) hypothesis cannot be disproved. In fact, because recrystallization of the original constituents of novaculite and colored chert has obliterated most original texture of the rocks, no hypothesis can be readily proved or disproved. Possible evidence to support the volcanic source hypothesis is the presence of terrigenous biotite and apatite, both common accessories in volcanic rocks, in turbidite layers in the formation. Apatite is widespread, but biotite was found in only one bed and is thus trivial. Much terrigenous quartz in the turbidites is nonundulose and free of many liquid or mineral inclusions, characteristics of quartz of volcanic origin (Krynine, 1946; Blatt and Christie, 1963). However, no grains with negative crystals, also typical of volcanic quartz, were found in either chert or turbidite beds. It can be concluded only that some detrital quartz may be of volcanic derivation. Metarhyolite pebbles from the boulder beds of the Haymond Formation of Pennsylvanian age offer evidence of mid-Paleozoic volcanism in the region. The pebbles in the boulder beds at the eastern edge of the Marathon Basin are derived from the landmass of Llanoria that contributed much siliciclastic debris to the geosyncline during late Paleozoic time (King, 1937; Cotera, 1962; McBride, 1966). Radiometric (Rb/Sr) dating of the pebbles (Denison and others, 1969) indicates a Siluro-Devonian age of volcanism and concomitant deformation, which coincides with the age of the lower part of the Caballos proposed here and earlier (Thomson, 1964, 1965). As emphasized by Denison and others (1969), this dating establishes that an episode of volcanism in Llanoria coincides with deposition of part of the Caballos, and suggests a probable source of some silica now present in the formation, but does not prove direct contribution of ash to the depositional basin.

Evidence against a direct volcanic origin is impressive. During this study only two grains of igneous rock fragments of volcanic or hypabyssal origin were found in turbidite layers, and no euhedral grains of zircon or apatite, common accessory minerals in volcanic rocks, were found, and no relict shard texture is detectable. This is a particularly low concentration of constituents of volcanic derivation (unless the quartz grains are volcanic); furthermore, the rock fragments may be older than the Caballos. Park (1961) and Park and Croneis (1969) report determinations of thorium and uranium concentrations made from one sample of Caballos novaculite and one sample of underlying Maravillas chert and concluded that the concentrations and Th/U ratio were not those typical of volcanic rocks. Thin dikes and sills from 1 to 3 mm wide of trachyte were found invading colored chert beds of the Caballos on East Bourland Mountain (Loc. 18) during this study. However, the intrusive layers postdate healed fractures in the chert and are clearly younger than the formation. Igneous plugs of the Marathon region are of Tertiary age (King, 1937). The absence of submarine flow rocks in the formation is a feature in common with many bedded chert formations (Grunau, 1965).

Evidence for Fan's hypotheses for the origin of the formation is not given in his abstract. From notes taken during the oral presentation of his paper at the annual meeting of the Geological Society of America, it appears that the petrified logs (from the upper chert and shale member) are the chief evidence for fluvial origin of his middle unit. A marine origin for these beds that are rich in Radiolaria is far more plausible, and the logs are best interpreted as objects that floated to their site of deposition. Diatoms are not known to occur in Paleozoic rocks (Siever, 1962), so that Fan's interpretation that the lower novaculite formed as diatomaceous ooze is untenable.

Bennett's suggestion that some inorganically (orthochemically) precipitated chert is present in the formation is difficult to evaluate. All available data on the behavior of silica in solution and in the ocean (Alexander and others, 1954; Bien and others, 1959; Fournier, 1960; Iler, 1955; Kennedy, 1950; Lier, 1959; Siever, 1957, 1962) indicate that silica will not spontaneously precipitate from sea water with a normal concentration of silica. The geochemical arguments against the inorganic origin of bedded chert are aptly treated by Siever (1962).

Bennett also suggested that some novaculite formed by the replacement of limestone because he found possible relict fecal pellets and mound structures that resemble those found in modern carbonate mud environments. However, the presence of pellets and mounds merely proves the sediment was unconsolidated; composition may not be inferred so easily. Pellets and mounds of organic and gas origin could have formed just as easily in ooze or mud of siliceous composition, as in carbonate. Because cherts in the Caballos that formed by the replacement of calcarenite beds have relict clastic texture and fabrics distinctly different from novaculite, it is highly unlikely that proto-novaculite formed by replacement of this type of limestone. In addition, the geologic setting of the Caballos is totally different from that known for either modern or ancient deposits of pure carbonate mud, and, thus, it is unlikely that proto-novaculite was originally such a deposit. The manganiferous carbonate grains present in novaculite and colored chert beds in small amounts is clearly of authigenic origin and probably postdates the microquartz.

Park and Croneis's intrepretation that the silica of the formation was precipitated as amorphous grains by organisms is the interpretation favored here; however, the idea that the grains were chiefly in the 0.1 to 5.0 μ size range is open to question.

The process whereby opal skeletons were converted to microquartz is unknown: whether by crystallization and dehydration ("devitrification" process of Baker and Bowman) or by solution and reprecipitation. The absence of shrinkage phenomena in the formation suggests the opal was destroyed before burial of more than a few feet, however. Electron micrographs show the novaculite is composed of many grains in the 0.1 to 5.0 μ size range mentioned by Park and Croneis, but it seems reasonable that these may be merely the size of crystals of microquartz formed originally by either method cited above.

Park and Croneis's interpretation that the sea floor was a reducing environment need not be true. As emphasized by Krumbein and Garrels (1952), a sediment surface may be an oxidizing environment (positive Eh), but reducing conditions may exist a few inches below the surface; and diagenetic events of the later reducing phase may be obvious to the exclusion of others. The following suggest the sea floor was oxidizing at times:

(1) Organic matter in the novaculite and many colored chert beds was completely destroyed. Although Radiolaria settle so slowly that they may not have had fleshy material left by the time they reached the sea floor, whatever organic material reached the site was destroyed by bacteria. These may have been anaerobic, but the intensity of destruction favors aerobic bacteria.

(2) Hematite in red shale is primary or early diagenetic and indicates oxidizing conditions during deposition; furthermore, it indicates that the diagenetic environment was not sufficiently euxinic to reduce it.

(3) The burrows in chert beds show organisms could survive below the sediment-water interface, a condition favored by mild oxygenation. The lack of preserved laminae in most novaculite and chert beds is probably also the result of bioturbation, adding weight to the significance of an infauna.

Special Problems

Depth of Water. The water depth during accumulation of Caballos sediment is a controversial topic; it has been interpreted to be both "shallow" and "deep." Evidence that can be cited as favoring either origin is summarized and criticized below.

EVIDENCE FAVORING A SHALLOW-WATER ORIGIN:

(1) The presence of fecal pellets and pit-and-mound structures in chert is compatible with the interpretation that the protonovaculite sediment was a carbonate mud similar to that accumulating in parts of Florida Bay and that novaculite formed by replacement. The present texture of much novaculite cannot disprove a carbonate-mud parent; but where relict texture is visible, it is everywhere that of spicules or Radiolaria. In addition, pits-and-mounds of organic origin are found today from intertidal to abyssal depths.

(2) The presence of limestone beds composed of fragments of shallow-water organisms. Although the detritus is clearly from shallow water, we believe the detritus was transported basinward by turbidity currents.

(3) Aberdeen (1940) favored a shallow-water origin because most Radiolaria genera from the Caballos are shallow-water dwellers today. The interpretation is not convincing, however, in view of the simultaneous occurrence in the Caballos of genera that today are found only in deep water. In pelagic sediments the deepest water forms are generally the best guides to ecology. In addition, the selective solution of certain forms of Radiolaria during diagenesis (Berger, 1968) has certainly yielded a different assemblage in chert samples than the original thanatocoenose.

(4) The most abundant identifiable skeletal particles in novaculite are spicules of probable sponges, and sponges are chiefly

shallow-water sessile organisms. However, many modern siliceous sponges live in deep water (de Laubenfels, 1955, p. E33); the ecology of Siluro-Devonian sponges is uncertain. Spicules of modern dead sponges are transported to deep water by marine currents.

EVIDENCE FAVORING A DEEP-WATER ORIGIN:

(1) The Caballos chert is similar to other bedded chert formations for which evidence of deep water is stronger. Mesozoic chert is locally overlain or underlain by marl rich in planktonic carbonate skeletons. The lack of carbonate skeletal grains in the chert in contrast to marl suggests the siliceous sediment accumulated below the compensation depth of carbonate, which, although highly variable, is commonly 5000 m in the modern ocean. Although Caballos chert is not associated with marl, it lacks planktonic carbonate skeletal grains; this may be due either to lack of supply of such grains or loss by solution because of depth. The presence of carbonate particles in the calcarenite turbidites suggests that carbonate was available at various times during accumulation of proto-chert, but whether such particles reached the basin is uncertain. Such particles may have been the source of the carbonate for the manganocalcite euhedra present in some Caballos chert beds.

(2) Holocene siliceous oozes are abyssal deposits. The analogy with Holocene ooze, however, is problematical because (a) Holocene ooze includes organisms not present in Siluro-Devonian time, (b) Holocene ooze is not as pure in silica as Caballos proto-chert must have been, and (c) Holocene ooze is accumulating in deep oceans on oceanic crust, whereas the Caballos and probably all similar ancient bedded-chert formations in linear geosynclines apparently accumulated on sialic crust. The latter point raises the question of whether modern environments are typical of or similar to those represented in ancient geosynclinal rocks—a question whose probable answer is no, but whose documentation is beyond the scope of the present problem. Woolnough (1942) presents a strong argument for the previous existence of broad shelves adjacent to peneplains and a situation where sediments like those forming today in deep water accumulated in moderately shallow water. From the evidence available, this hypothesis can neither be accepted nor rejected.

(3) Sedimentary features of distinctly shallow-water origin are absent.

(4) Terrigenous grains are scarce in the formation; there is an exceptionally high ratio of organic to terrigenous particles.

These features suggest deposition far from shore, where water depths are generally great. Similar conditions could also be explained by the Woolnough hypothesis, however.

(5) Shallow-water infauna and epifauna in place are absent. Trace fossils present are of burrowers and tunnelers, presumably worms, because no hard parts have been found. Trace fossils in alleged deep-water flysch deposits are characterized by lateral feeding, crawling, and burrow marks (Seilacher, 1958); hence, the vertical burrows in the Caballos are anomalous to a deep-water interpretation. Other evidences, however, indicate that the sea floor was generally oxidizing in contrast to common deep-water muddy environments that are reducing, and may account for the desire of presumed deep-water infauna to burrow into beds.

(6) Bedding structures and sediment type indicate deposition occurred below surf-base and probably below wave-base. However, surf-base may be as shallow as 30 feet (Dietz, 1963), and wave-base varies with the grain size under consideration (Moore and Curray, 1964). Considering the grain size of Caballos siliceous ooze, it is likely that wave-base was several hundred feet deep.

(7) The formation is underlain and overlain by what we (Thomson, 1964, 1965; Thomson and McBride, 1964; McBride, 1968a, 1968b) interpret to be deep-water deposits without an interruption in sedimentation. However, our depth interpretation for the subjacent formations is based on sedimentological evidence which cannot be verified by independent data.

In summary, we conclude that conclusive evidence for depth of water during accumulation of Caballos sediment is lacking. When all data are considered, however, we believe that the deep-water interpretation strongly outweighs the shallow-water interpretation and that the depth of water was probably greater than 300 feet and may have been several thousand feet deep. This interpretation must be revised if further support for the Woolnough hypothesis is forthcoming.

Purity of Novaculite. The near absence of organic matter and grains of terrigenous clay in the novaculite are the chief reasons for its lack of color. Obviously whatever organic matter reached the sea floor was subsequently destroyed. If the large pit-and-mounds in the novaculite are of gas origin, this attests to emission of probably organically derived gas generated below the sediment interface. Apparently not all organic matter was destroyed at the interface.

The lack of clay in the novaculite (except for rare partings) suggests a sea of remarkable clarity, and yields several implications concerning paleogeography (*see* below).

The intertonguing of novaculite and colored chert along the southern edge of the Dagger Flat anticlinorium suggests that novaculite accumulated contemporaneous with colored chert during part of Caballos deposition. Thus, the environmental controls that produced novaculite instead of colored chert were those that operated at the bottom of the basin and not vertically through the water column. By equating thickness of novaculite beds (and members) to duration of accumulation, the conclusion follows that bottom conditions were uniform for incredibly long periods of time in large parts of the basin while proto-novaculite accumulated. These conditions are unmatched today by environments known in deep-oceanic or other marine environments.

Red *versus* Green Shale. Red shale is a conspicuous, but minor part of the formation, and contrasts strongly with the predominant green shale and chert of the Caballos. This association of red and green beds in bedded chert sequences is a common one, the origin of which is not understood (Grunau, 1959, p. 116-137; 1965, p. 196-197). Chemical analyses of red and green shale samples from the Caballos (Table 3) show the samples have the same amount of FeO, but the red shale has nearly 5 percent more Fe_2O_3 than the green shale. Illite with ferrous iron is apparently the cause of the green color in chert and shale in the Caballos. The red shale is colored by hematite, whose stratigraphic distribution and disseminated occurrence suggests that it (or the original iron-bearing phase) was a primary constituent of the shale upon deposition and subsequently has not been reduced. Red shale locally has been bleached to green shale along joints, but the widespread occurrence of green chert and shale show that the green color is certainly a primary feature.

Grunau (1965, p. 196-197) infers that the hematite in red chert in the Tethys geosyncline was derived from red upland soils and was preserved by oxidizing conditions in the depositional basins. It is uncertain how the original iron phase accumulated in the Caballos sediment, but the composition of bottom water probably controlled whether iron oxide pigmented red shale, or unpigmented (green) shale accumulated. Magnetite spheres of presumably meteoritic origin have not been found in the Caballos, although a meteoritic source for some hematite cannot be ruled out. Caballos red shale is probably analogous to red or brown clay of modern deep-ocean basins. It is similar in chemical composition to deep-sea red clay (Clarke, 1924, p. 518), except that the K_2O/Na_2O ratio of the Caballos shale is 8 and that for red clay is 1. The value for the Caballos is similar to that for red shale in the Franciscan Group (Bailey and others, 1964).

Rate of Deposition. That the Caballos represents a relatively slow rate of deposition has been emphasized by several workers (King, *in* Flawn and others, 1961, p. 180; Goldstein, 1959; Flawn, 1964, p. 11; Thomson, 1964). One of us (Thomson, 1964, p. 15) has previously calculated a minimum rate of uncompacted sediment accumulation of 0.46 cm/1000 years for the Caballos. The sediment was estimated to have compacted to one-half its original thickness, and it was assumed that the duration of deposition was at least all of Silurian and Devonian time. A value of 600 ft. was used as the present maximum thickness of the Caballos, and 8×10^7 years as the duration of deposition. This computed value of 0.46 cm/1000 years gives the magnitude of the minimum rate of accumulation for the thickest part of the formation. Additional computations based on the maximum and minimum thickness values determined from this study and using the previous compaction and duration of deposition figures yield values of 0.53 cm/1000 years for the maximum rate of deposition and 0.10 cm/1000 years for the minimum rate. Table 7 gives the calculated rates of deposition of some Holocene oceanic and other marine sediments. The Caballos

TABLE 7. RATE OF ACCUMULATION OF MARINE SEDIMENTS IN CM/1000 YEARS ASSUMING ZERO POROSITY (*after* Kuenen, 1967, p. 13)

Recent pelagic clay	0.05
Recent blue mud	0.5
Globigerina ooze	0.5
Turbidites on abyssal plains	4
Basins off California	13
Adriatic Sea	5
Black Sea	4
Blake Plateau	0.2
16 North American geosynclines, maximum over long periods	7.5
23 World geosynclines, maximum over short periods	20
Estimate for whole geosynclines between orogenies	2.5

rates range from slightly lower to greatly lower than the values for the Holocene oceanic sediments, with the exception of Holocene pelagic clay, and much lower than the average rate of geosynclinal accumulation in the past. The Caballos rates of deposition are even less than shown if the formation also includes Early Mississippian time. The assumption has been made that the Caballos is a time-stratigraphic unit, and although this is a reasonable assumption, it cannot be documented at present (*see* Part 1). It is also possible that the original siliceous ooze may have "condensed" much more than to 50 percent its original thickness because solution and reprecipitation of the siliceous particles took place in addition to general compaction.

Spicule Orientation. The presence of spicules in novaculite that have a strong preferred orientation attests to the existence of currents that swept the sea floor during the accumulation of Caballos sediment. The strength, frequency, and duration of such currents are unknown, but probably the currents were capable of transporting sufficient skeletal particles to smooth up sea-floor irregularities that are visible today as uneven bedding planes, and possibly of winnowing clay from biogenic remains to produce clay-rich and silica-rich layers. The existence of bottom currents in the modern deep oceans is well documented (Swallow and Worthington, 1957; Stride, 1963), but little is known about their duration, velocity, or effectiveness in transporting sediment.

Chemical Diagenesis. The process whereby opal skeletons converted to microquartz to produce Caballos chert is unknown, but three mechanisms are suggested: (a) solution of opal and reprecipitation of microquartz from pore solutions, (b) crystallization and dehydration (devitrification process of Baker and Bowman) of opal tests, and (c) conversion of opal tests to silica gel that subsequently dewatered and crystallized.

The present compact and nonporous fabric of chert, information on chemistry of silica, and study of recent oceanic sediments strongly suggest that mechanism (a) above is the chief process of chertification. According to Folk and Weaver (1952), chalcedony is generally a cavity-filling form of quartz. The occurrence of chalcedony pseudomorphs of Radiolaria in the Caballos, although not common, suggests this form of quartz precipitated in cavities produced by solution of opal tests. From a study of Holocene siliceous oceanic sediments Arrhenius (1963, p. 692-3) states:

The instability of silica in the interstitial water of the sediment causes continuous dissolution of the siliceous fossils after deposition: the silicoflagellates disappear first, followed by diatoms, then radiolarians, and finally even robust sponge spicules. . . . In some cases reprecipitation of the silica as opal takes place, resulting in thin flakes of this mineral along bedding planes in phosphorite, or in laminae of chert, observed to occur at depth in pelagic sediments in a few instances.

Thus, it is envisaged that solution of the finest and most delicate opal tests was followed by reprecipitation of silica as a cement around more robust forms to ultimately produce a compact rock.

The skeletal grains whose morphology is preserved in the chert may have converted directly to a mosaic of microquartz, process (b) stated above.

Under hydrothermal conditions amorphous silica converts to low-quartz through the intermediate stage of low-cristobalite, and a similar transformation is suggested in chert formation (Mizutani, 1966). In chert as old as the Caballos, however, low quartz is the only phase present (Mizutani, 1966, Fig. 12), and previous phases can only be inferred.

There is no evidence to support the formation of Caballos chert by mechanism (c) above, although phase transformations similar to this one were proposed for the origin of bedded chert in California by Davis (1918), Taliaferro (1934) and Bailey and others (1964). These authors argued that colloidal silica flocculated inorganically from sea water and passed from the sol stage (Davis, Taliaferro) through gels into microquartz. Except for possibly two samples, objects described by Taliaferro (1934) as contraction spheroids and cited as evidence of an original gel state of proto-chert have not been found in the Caballos.

Davis and Taliaferro supported the idea that in the Franciscan Group, clay was expelled from an original clayey gel mixture during diagenesis to form distinct chert and shale beds. Textures in the Caballos do not support such a mechanism. However, lenses of pure chert from 1 to 2 cm thick locally occur in siliceous shale host rock and together have slightly contorted bedding. Textures in thin section (Pl. 14A) suggest that the lenses of chert grew by local solution of material of a thin siliceous layer and subsequent precipitation of the silica at selected positions along that layer. That is, silica was largely removed along part of the original layer of ooze, whereas elsewhere the silica was reprecipitated to thicken the layer and form the lenses. The contorted bedding formed by a combination of lens growth and differential compaction.

The different grain sizes of microquartz in the chert beds may be a product of either (a) an increase in grain size of originally UFG microquartz particles with time, or (b) original differences in grain size determined by the spacing of centers of crystallization of opal, or (c) combination of both the above processes. Present textures in different rocks favor one interpretation over the other, but evidence to define the major process is lacking. Slump-formed fractures in colored chert beds are generally filled with chalcedony, but a few are filled by UFG and VFG microquartz. Thus, the youngest microquartz is the finest grained. Probably the specks of UFG and VFG microquartz in novaculite were the last (most recent) Radiolaria to undergo crystallization from opal.

Beds rich in manganese minerals are common in bedded chert formations and have been mined from the Arkansas Novaculite (Miser, 1917) and Franciscan-Knoxville group (Taliaferro and Hudson, 1943), but none have been mined profitably from the Caballos. The above authors interpret the manganese to have been deposited from sea water as a primary constituent of proto-chert, but the manganese has been segregated and enriched in fractures and pods by secondary processes. All manganese occurrences in the Caballos are secondary segregations, with the possible exception of minutely dispersed irresolvable particles in chert. The only manganese rich mineral phase identified in the Caballos is pyrolusite. Pyrolusite occurs in Holocene oceanic manganese nodules (Buser and Grutter, 1956), but is more common as a secondary manganese mineral.

The hematite pigment in Caballos shale is inferred to be a primary deposit because of its stratigraphic distribution. Iron in Holocene manganese nodules is precipitated either as goethite or disordered FeO(OH) (Arrhenius and Korbisch, 1959; Arrhenius, 1963, p. 666-7), and a similar primary phase for the iron in the Caballos is likely. Hematite in Caballos chert has been diagenetically redistributed along bedding planes and fractures; it also occurs as an alteration of pyrite in nodules and, locally, in chert as pigment.

PALEOGEOGRAPHY AND PROVENANCE

Regional paleogeography during deposition of the Caballos is poorly known because rocks of equivalent age do not crop out in surrounding areas, and subsurface information is spotty except to the northwest. The picture is complicated by the uncertainty of how far the present outcrops have moved from their original position during Pennsylvanian thrusting. The Gulf Oil Company's No. 1 D. S. Combs well 1 mile southeast of Marathon, and the Slick-Urschel Oil Company's No. 1 Decie-Sinclair well, 3 miles northwest of Marathon, both penetrated rocks of the foreland or crantonic facies beneath deformed rocks of the Marathon (or Ouachita) facies (Wilson, 1954; Flawn and others, 1961, p. 234-238; Ross, 1962). Hence, the Marathon facies is clearly allochthonous along the western part of the basin. Wells to the east in the basin do not pass through rocks of Marathon facies.

The distribution of coarse clastic beds (calcarenite, sandstone, conglomerate) in the Caballos in the Marathon Basin indicates the detritus was derived chiefly from a source to the northwest. A positive element of low relief existed to the west throughout most of Ordovician time (Wilson, 1954; McBride, 1967, 1968a; Young, 1968) and was probably peneplaned by the start of Caballos deposition. Much of the granule and gravel size detritus was derived cannibalistically from Maravillas and Caballos chert beds when the western margin of the basin was uplifted. Quartz was probably derived from older Cambrian and Ordovician strata, whereas the calcarenite debris was shed from a carbonate platform that was contemporaneous with the basinal biogenic silica deposits.

Clay may have been intermittently derived from Llanoria to the east; this is the probable source of clay in the upper part of the Caballos and in beds transitional into the overlaying Tesnus Formation.

Conditions of deposition probably were uniform from the Solitario to the Marathon Basin because rocks are similar in character in these regions. However, conditions were not uniform between the Marathon Basin and the Oklahoma-Arkansas part of the Ouachita geosyncline because only a dark chert facies has been recovered from deep wells between the areas (Flawn and others, 1961).

Regional subsurface data show that a large marine basin, the Tabosa Basin, extended northwest from the Marathon geosyncline during early Paleozoic time (McGlasson, 1967) and covered part of west Texas and southeastern New Mexico. According to McGlasson (1967), rocks correlative with the Caballos in the Tabosa Basin include four relatively deep-water stratigraphic units:

Fusselman Formation (limestone and dolomite), "Upper Silurian" unit (shale, limestone, and dolomite), "Devonian" unit (chert and siliceous limestone), and the Woodford Formation (dark shale). Here also, only very fine clastic material, in addition to carbonate, reached the depositional basin.

GEOLOGIC HISTORY

The initiation of deposition of nearly pure siliceous ooze of the Caballos Formation began when the influx of clay and organic matter that characterize the uppermost sediments of the Maravillas Chert ceased. The Maravillas Chert is interbedded fetid turbidite calcarenite, pelagic spiculitic chert and shale, and minor conglomerate and replacement dolomite that accumulated in deep water (Thomson and McBride, 1964; McBride, 1968a); the formation is latest Ordovician (Berry, 1960). Caballos deposition followed the Maravillas without interruption, and is inferred to have begun in latest Ordovician or early Silurian time.

Uniform conditions existed throughout the depositional basin during accumulation of the lower chert member (up to 15 feet thick). Small amounts of silt and clay were brought to the basin by winds, storms, and turbidity currents to make up 2 to 5 percent of the resulting tan chert beds. When the influx of terrigenous material essentially ceased, nearly pure biogenic ooze that makes up the lower novaculite member (up to 150 ft. thick) accumulated. All organic matter that reached the depositional interface was oxidized; all carbonate was dissolved, and only traces of iron-oxides were present. The biogenic material was largely capsules of Radiolaria that lived in surface water and opal spicules from sponges that lived at the margin of the basin and whose debris was moved into deep water by marine currents.

Weak tectonic disturbance either rejuvenated the peneplaned source area or changed the marine current pattern so that considerable clay and minor silt was added to the constant accumulation of siliceous ooze to provide the interbedded colored chert and shale beds of the lower chert and shale member (up to 200 feet thick). This tectonic episode probably triggered the turbidity currents that deposited sandstone beds at the base of this member, although several turbidite calcarenite beds occur higher in the section also. Radiolaria are more conspicuous than spicules in the colored chert beds, but whether this represents originally greater abundances of Radiolaria than spicules or selective obliteration of spicules during diagenesis is unknown.

A second cessation of influx of terrigenous clay and silt initiated the accumulation of the upper novaculite member (up to 400 feet thick), and the site of maximum basin subsidence was east of that during previous deposition. Additional tectonism again resulted in the accumulation of colored proto-chert and mud beds of the upper chert and shale member (up to 250 feet thick). During this interval of deposition, earthquakes triggered several episodes of submarine slumping during which weakly lithified chert beds were contorted or brecciated. Other significant events included accumulation of several tens of feet of ooze-free red clay, settling of logs of *Callixylon* that floated out to sea, and deposition of manganese that was subsequently concentrated during early diagenesis.

Uplift in Llanoria to the southeast caused an increasing rate of mud accumulation that gradually diluted the siliceous oozes; ooze beds diminished upsection and were succeeded by black siliceous shale and finally by black nonsiliceous shale of the Tesnus Formation (Late Mississippian to Early Pennsylvanian age). Caballos deposition is inferred to have continued through Devonian time and probably into Mississippian time.

Siliceous ooze of the Caballos converted to chert during early diagenesis; the diagenesis was implemented by lithostatic load.

References Cited

Aberdeen, Esther, 1940, Radiolarian fauna of the Caballos Formation, Marathon Basin, Texas: Jour. Paleontology, v. 14, p. 127-139.

Alexander, G. B., Heston, W. M., and Iler, R. K., 1954, The solubility of amorphous silica in water: Jour. Phys. Chemistry, v. 58, p. 453-455.

Arick, M. B., 1935, Early Paleozoic unconformities in Trans-Pecos Texas (Cambrian to Devonian inclusive): Texas Univ. Bull. 3501, p. 117-123.

Arrhenius, G., 1963, Pelagic sediments, p. 655-727, *in* Hill, M. N., *Editor*, The Sea: New York, Interscience Publishers, 963 p.

Arrhenius, G., and Korbisch, H., 1959, Uranium and thorium in marine minerals, p. 497 *in* Sears, M., *Editor*, Internatl. Oceanog. Cong. Preprints: Washington, D.C., Am. Assoc. Adv. Sci. Pub.

Aubuoin, Jean, 1965, Geosynclines: New York, American Elsevier Publishing Co., Inc., 335 p.

Baker, C. L., and Bowman, W. F., 1917, Geologic exploration of the southeastern Front Range of Trans-Pecos Texas: Texas Univ. Bull. 1753, p. 61-177.

Bailey, E. H., Irwin, W. P., and Jones, D. L., 1964, Franciscan and related rocks and their significance in the geology of western California: California Div. Mines and Geol. Bull. 183, 177 p.

Barghoorn, E. S., and Tyler, S. A., 1965, Microorganisms from the Gunflint Chert: Science, v. 147, no. 3658, p. 563-577.

Bennett, R. E., 1959, Geology of East Bourland and Simpson Springs Mountains, Brewster County, Texas: M.A. Thesis, Univ. of Texas, Austin, Texas, 172 p. (Available by interlibrary loan.)

Berger, W. H., 1968, Radiolarian skeletons: solution at depths: Science, v. 159, no. 3820, p. 1237-1239.

Berry, W. B. N., 1960, Graptolite faunas of the Marathon region, west Texas: Texas Univ. Pub. 6005, 170 p.

Berry, W. B. N., and Nielsen, H. M., 1958, Revision of Caballos Novaculite in Marathon region, Texas: Am. Assoc. Petroleum Geologists Bull., v. 42, p. 2254-2259.

Bien, G. S., Contois, D. E., and Thomas, N. H., 1959, The removal of soluble silica from fresh water entering the sea, *in* Silica in sediments: Soc. Econ. Paleontologists and Mineralogists Spec. Pub. 7, p. 20-35.

Bjorklund, T. K., 1962, Structure of Horse Mountain anticline (southwest extension), Brewster County, Texas: M.A. Thesis, Univ. of Texas, Austin, Texas, 74 p. (Available by interlibrary loan.)

Blatt, Harvey, and Christie, J., 1963, Undulatory extinction in quartz of igneous and metamorphic rocks and its significance in provenance studies of sedimentary rocks: Jour. Sed. Petrology, v. 33, p. 559-579.

Buser, W., and Grutter, A., 1956, Untersuchungen and Mangansedimenten: Schweizer. Mineralog. u. Petrog. Mitt., v. 36, p. 49-62.

Byrd, W. M., 1958, The geology of a portion of the Combs Ranch, Brewster County, Texas: M.A. Thesis, Univ. of Texas, Austin, Texas, 67 p. (Available by interlibrary loan.)

Calvert, S. E., 1966, Accumulation of diatomaceous silica in the sediments of the Gulf of California: Geol. Soc. America Bull., v. 77, p. 569-596.

Carver, R. E., 1965, Undulose extinction in fine-grained rocks: Jour. Sed. Petrology, v. 35, p. 980-983.

Clarke, F. W., 1924, The data of geochemistry, 5th edition: U.S. Geol. Survey Bull. 710, 841 p.

Comstock, T. B., 1888, Report upon preliminary examination of the geology of western central Arkansas: Arkansas Geol. Survey Ann. Rept., v. 1, 320 p.

Cotera, A. S., 1962, Petrology and petrography of Mississippian-Pennsylvanian Tesnus formation, Marathon Basin, Trans-Pecos Texas: Ph.D. Dissert., Univ. of Texas, Austin, Texas, 186 p. (Available by interlibrary loan.)

Daly, Jesse, 1964, Application of modern geosyncline concepts to the Marathon region of west Texas, *in* The filling of the Marathon geosyncline: Soc. Econ. Paleontologists and Mineralogists Permian Basin Sec., Pub. 64-9 p. 47-51.

Davis, E. F., 1918, The Radiolarian cherts of the Franciscan Group: California Univ. Dept. Geol. Bull., v. 11, no. 3, p. 235-432.

de Laubenfels, M. W., 1955, Porifera: Treatise on invertebrate paleontology, Part E: Geol. Soc. America, p. E21-E112.

Denison, R. E., Kenny, G. S., Burke, W. H., Jr., and Hetherington, E. A., Jr., 1969, Isotopic ages of igneous and metamorphic boulders from the Haymond Formation (Pennsylvanian), Marathon Basin, Texas, and their significance: Geol. Soc. America Bull., v. 80, p. 245-256.

Derby, O. A., 1898, Notes on Arkansas novaculite: Jour. Geology, v. 6, p. 366-368.

REFERENCES CITED

Dietz, R. S., 1963, Wave-base, marine profile of equilibrium, and wave-built terraces: a critical appraisal: Geol. Soc. America Bull., v. 74, p. 971-990.

Eifler, G. K., 1943, Geology of the Santiago Peak Quadrangle, Texas: Geol. Soc. America Bull., v. 54, p. 1613-1644.

Ellison, S. P., 1962, Conodonts from Trans-Pecos Paleozoic of Texas (Abs.): Am. Assoc. Petroleum Geologists Bull., v. 46, p. 266.

Eugster, H. P., 1967, Hydrous sodium silicates from Lake Magadi, Kenya: precursors of bedded chert: Science, v. 157, no. 3793, p. 1177-1180.

Fagin, J. J., 1963, Carboniferous cherts, turbidites, and volcanic rocks in Northern Independence Range, Nevada: Geol. Soc. America Bull., v. 73, p. 595-612.

Fan, P. H., 1964, Revision of the Caballos Novaculite of Trans-Pecos Texas (Abs.): Geol. Soc. America Spec. Paper 76, p. 56-57.

Featherstonhaugh, G. W., 1835, Geological report of an examination made in 1834 of the elevated country between the Mississippi and Red Rivers: Washington, D.C., 97 p.

Flawn, P. T., 1964, The regional setting of the Marathon salient, *in* The filling of the Marathon Geosyncline: Soc. Econ. Paleontologists and Mineralogists Permian Basin Sec. Pub. 64-9, p. 9-11.

Flawn, P. T., Goldstein, August, Jr., King, P. B., and Weaver, C. E., 1961, The Ouachita system: Texas Univ. Bull. 6120, 401 p.

Folk, R. L., 1965, Petrology of sedimentary rocks: Austin, Hemphills, 159 p.

Folk, R. L., and Weaver, C. E., 1952, A study in the texture and composition of chert: Am. Jour. Sci., v. 250, p. 498-510.

Fournier, R. O., 1960, Solubility of quartz in water in the temperature interval from $25°$ C to $300°$ C: Geol. Soc. America Bull., v. 71, p. 1867-1868.

Goldstein, August, Jr., 1959, Cherts and novaculites of the Ouachita facies, *in* Ireland H. A., *Editor,* Silica in sediments: Soc. Econ. Paleontologists and Mineralogists Spec. Pub. 7, 185 p.

Goldstein, August, Jr., and Hendricks, T. A., 1953, Siliceous sediments of Ouachita facies in Oklahoma: Geol. Soc. America Bull., v. 64, p. 421-442.

Govett, G. J. S., 1966, Origin of banded iron formations: Geol. Soc. America Bull., v. 77, p. 1191-1212.

Graves, R. W., Jr., 1952, Devonian conodonts from the Caballos Novaculite: Jour. Paleontology, v. 26, p. 610-612.

―――― 1954, Geology of the Hood Springs Quadrangle, Brewster County, Texas: Texas Univ. Bur. Econ. Geology Rept. Inv., no. 21, 51 p.

Griswold, L. S., 1892, Whetstones and the novaculites of Arkansas: Arkansas Geol. Survey Ann. Rept. (1890), v. 3, 443 p.

Grunau, H. R., 1959, Mikrofacies und Schichtung ansgewahlter, jungmesozoischer, Radiolarit-fuhrender Sedimentserien der Zentral-Alpei: Internat. Sedimentary Petrography, ser. 4, Leiden, E. J. Brill, 179 p.

—— 1965, Radiolarian cherts and associated rocks in space and time: Eclogae Geol. Helvetiae, v. 58, p. 157-208.

Harlton, B. H., 1953, Ouachita chert facies, southeastern Oklahoma: Am. Assoc. Petroleum Geologists Bull., v. 37, p. 778-796.

Hazzard, R. T., Maxwell, R. A., and Lonsdale, J. T., 1958, Paleozoic rock exposures, Persimmon Gap — Dog Canyon areas, Brewster County, Texas: Am. Assoc. Petroleum Geologists Bull., v. 42, p. 887.

Henbest, L. G., 1936, Radiolaria in the Arkansas novaculite, Caballos novaculite and Bigfork chert: Jour. Paleontology, v. 10, p. 76-78.

Hendricks, T. A., Knechtel, M. N., and Bridge, Josiah, 1937, Geology of Black Knob Ridge, Oklahoma: Am. Assoc. Petroleum Geologists Bull., v. 21, p. 1-26.

Herrin, E. T., 1958, Geology of the Solitario area, Trans-Pecos Texas: Ph.D. Dissert., Harvard Univ., Cambridge, Massachusetts, 183 p.

Hill, R. T., 1900, Physical geography of the Texas region: U.S. Geol. Survey Topographic Atlas, Folio 3, 12 p.

Honess, C. W., 1923, Geology of the Southern Ouachita Mountains of Oklahoma: Oklahoma Geol. Survey Bull. 32, Pt. I, 278 p.

Hough, J. L., 1958, Fresh-water environment of deposition of Precambrian iron formations: Jour. Sed. Petrology, v. 28, p. 414-430.

Iler, R. K., 1955, The colloidal chemistry of silica and silicates: Ithaca, Cornell Univ. Press, 324 p.

Jones, T. S., 1953, Stratigraphy of the Permian basin of west Texas: Midland, West Texas Geol. Society Pub. 57. p.

Kaibara, Hisashi, 1964, A study of the micro-texture of cherts: Kyoto Univ. Coll. Sci. Mem., ser. B, v. 30, no. 4, p. 59-73.

Keller, W. D., 1953, Illite and montmorillonite in green sedimentary rocks: Jour. Sed. Petrology, v. 23, p. 3-9.

Kennedy, G. C., 1950, A portion of the system silica-water: Econ. Geology, v. 45, p. 629-653.

King, P. B., 1930, The geology of the Glass Mountains, Texas: Texas Univ. Bull. 3038, 167 p.

—— 1931, Pre-Carboniferous stratigraphy of the Marathon uplift: Am. Assoc. Petroleum Geologists Bull., v. 15, p. 1059-1085.

—— 1937, Geology of the Marathon Region, Texas: U.S. Geological Survey Prof. Paper 187, 148 p.

Kirchmayer, Martin, 1959, Beitrage zur Frage des Cherts (= Feuersteins) I, Uber ein Geopetal gefuge im Chert (Flint) von Nevada, U.S.A.: Neues Jahrb. Geologie u. Paläontologie Abh., v. 5, p. 209-229.

REFERENCES CITED

Krumbein, W. C., and Garrels, R. M., 1952, Origin and classification of chemical sediments in terms of pH and oxidation-reduction potentials: Jour. Geology, v. 60, p. 1-33.

Krynine, P. D., 1946, Microscopic morphology of quartz types, Anales Sequndo Congr. Panamericano de Ing. de Minas y Geol., v. 3, p. 35-49 (publ. in 1950).

Kuenen, Ph. H., 1967, Geosynclinal sedimentation: Geol. Rundschau, v. 56, p. 1-18.

LaBerge, G. L., 1967, Microfossils and Precambrian iron-formations: Geol. Soc. America Bull., v. 78, p. 331-342.

Maxwell, R. A., 1949, Marathon region, Big Bend region, Green Valley-Paradise Valley region, Sierra Blanca region: West Texas Geol. Soc. Guidebook, 111 p.

Maxwell, R. A., and Hazzard, R. T., 1967, Stratigraphy, p. 26-156, *in* Geology of Big Bend National Park, Brewster County, Texas: Texas Univ. Pub., no. 6711, 320 p.

Maxwell, R. A., Lonsdale, J. T., Hazzard, R. T., and Wilson, J. A., 1955, Spring Field Trip: West Texas Geol. Soc. Guidebook 142 p.

McBride, E. F., 1966, Sedimentary petrology and history of the Haymond Formation (Pennsylvanian), Marathon Basin, Texas: Texas Univ. Bur. Econ. Geology Rept. Inv. 57, 101 p.

—— 1967, Sedimentology of the Woods Hollow Formation (Ordovician), Marathon region, Texas (abs.): Texas Jour. Science, v. 19, p. 417.

—— 1968a, Sedimentology of the Maravillas Formation (Ordovician), Marathon Basin, Texas (abs.): Geol. Soc. America Spec. Paper 115, p. 373.

—— 1968b, Origin of the Caballos Novaculite, Marathon Region, Texas (Abs.): Geol. Soc. America Spec. Paper 115, p. 141-142.

McBride, E. F., and Thomson, Alan, in press, Stratigraphy and origin of the Caballos Novaculite: Guidebook, Dallas Geol. Society.

McGlasson, E. H., 1967, The Siluro-Devonian of west Texas and southeast New Mexico: Tulsa Geol. Soc. Digest, v. 35, p. 148-164.

Miser, H. D., 1917, Manganese deposits of the Caddo Gap and DeQueen quadrangles, Arkansas: U.S. Geol. Survey Bull. 660C, p. 59-122.

Miser, H. D., and Purdue, A. H., 1929, Geology of the DeQueen and Caddo Gap quadrangles, Arkansas: U.S. Geol. Survey Bull. 808, 195 p.

Mizutani, Shinjiro, 1966, Transformation of silica under hydrothermal conditions: Nagoya Univ. Jour. Earth Sci., v. 14, p. 56-88.

Moore, D. G., and Curray, J. R., 1964, Wave-base, marine profile of equilibrium, and wave-built terraces: Discussion: Geol. Soc. America Bull., v. 75, p. 1267-1273.

Owen, D. D., 1860, Second report of a geological reconnaissance of the middle and southern counties of Arkansas: Philadelphia (*non vide*).

Park, D. E. Jr., 1961, The origin of bedded silicates with particular reference to the Caballos and Arkansas Novaculite formations: unpubl. Ph.D. dissert., Rice Univ., 82 p.

Park, D. E., Jr., and Croneis, Carey, 1969, Origin of the Caballos and Arkansas Novaculite formations: Am. Assoc. Petroleum Geologists Bull., v. 53, p. 94-111.

Pettijohn, F. J., 1957, Sedimentary rocks: Harper, New York, 718 p.

Powers, Sidney, 1921, Solitario Uplift, Presidio, and Brewster Counties, Texas: Geol. Soc. America Bull., v. 32, p. 417-428.

Ross, C. A., 1962, Permian tectonic history in Glass Mountains, Texas: Am. Assoc. Petroleum Geologists Bull., v. 46, p. 1728-1746.

Rutley, Frank, 1894, On the origin of certain novaculites and quartzites: Geol. Soc. London Quart. Jour., v. 50, p. 377-394.

Schoolcraft, H. R., 1819, A review of the lead mines of Missouri, including some observations on the mineralogy, geology, geography, antiquities, soil, climate, population, and production of Missouri, Arkansas, and other sections of the western country: New York, Wilson and Co., 299 p.

Seilacher, Adolf, 1958, Zur okologischen Charakteristik von Flysch und Molasse: Eclogae Geologicae Helvetiae, v. 51, p. 1062-1078.

Sellards, E. H., 1933, The pre-Paleozoic and Paleozoic systems in Texas, *in* The geology of Texas, Vol. 1, Stratigraphy: Texas Univ. Bull. 3232, p. 15-238.

Sellards, E. H., Adkins, W. S., and Arick, M. B., 1930, Geologic map of the Solitario of Texas: Texas Univ. Bur. Econ. Geology. Revised in 1931.

Siever, Raymond, 1957, The silica budget in the sedimentary cycle: Am. Mineralogist 42, p. 821-841.

—— 1962, Silica solubility 0-200° C and the diagenesis of siliceous sediments: Jour. Geology, v. 61, p. 127-149.

Sloss, L. L., 1963, Sequences in the cratonic interior of North America: Geol. Soc. America Bull., v. 74, p. 93-114.

Sterling, P. J., Stone, C. G., and Holbrook, D. F., 1966, General geology of eastern Ouachita Mountains, Arkansas, *in* Field conference on flysch facies and structure of the Ouachita Mountains: Kansas Geological Society, p. 177-194.

Stride, A. H., 1963, Current-swept sea floor near the southern half of Great Britain: Geol. Soc. London Quart. Jour., v. 119, p. 175-199.

Swallow, J. C., and Worthington, L. V., 1957, Measurements of deep currents in the western North Atlantic: Nature, v. 179, p. 1183-1184.

Taliaferro, N. L., 1934, Contraction spheroids in cherts: Geol. Soc. America Bull., v. 45, p. 189-232.

Taliaferro, N. L., and Hudson, F. S., 1943, Genesis of the manganese deposits of the Coast Ranges of California, *in* Manganese in California: Calif. Div. Mines Bull. 125, p. 217-275.

Thomson, Alan, 1964, Genesis and bathymetric significance of the Caballos Novaculite, Marathon Region, Texas, *in* The filling of the Marathon geosyncline, Symposium and Guidebook: Soc. Econ. Paleontologists and Mineralogists Permian Basin Sec. Pub. 64-9, p. 12-16.

—— 1965, Condition of deposition of the Caballos Novaculite, Marathon Region, Texas, (abs.): Geol. Soc. America Spec. Paper 82, p. 206.

Thomson, Alan, and McBride, E. F., 1964, Summary of the geologic history of the Marathon geosyncline, *in* The filling of the Marathon geosyncline, Symposium and Guidebook: Soc. Econ. Paleontologists and Mineralogists Permian Basin Sec. Pub. 64-9, p. 52-60.

Udden, J. A., 1907, A sketch of the geology of the Chisos country, Brewster County, Texas: Texas Univ. Bull. 93, 101 p.

Udden, J. A., Baker, C. L., and Bose, Emil, 1916, Review of the geology of Texas: Texas Univ. Bull. 1644, 178 p.

Van Hise, C. R., 1904, A treatise on metamorphism: U.S. Geol. Survey Monograph 47, p. 853.

van Lier, J. A., 1959, The solubility of quartz: Utrecht Kemink en Zoon, 54 p.

van Waterschoot, van der Gracht, W. A. J. M., 1931, Permo-Carboniferous orogeny in south-central United States: Am. Assoc. Petroleum Geologists Bull., v. 15, p. 991-1057.

von Streeruwitz, W. H., 1891, Report on the geology and mineral resources of Trans-Pecos Texas: Texas Geol. Survey, and Ann. Report (1890), p. 669-738.

Weed, W. H., 1902, Geological sketch of the Hot Springs district, Arkansas: 57th Congress, 1st session, Senate Decument 282, p. 84.

Wheeler, H. E., 1963, Post-Sauk and pre-Absaroka Paleozoic stratigraphic patterns in North America: Am. Assoc. Petroleum Geologists Bull., v. 47, p. 1497-1526.

Wilson, J. L., 1954, Ordovician stratigraphy in the Marathon folded belt, west Texas: Am. Assoc. Petroleum Geologists Bull., v. 38, p. 2455-2475.

Woolnough, W. G., 1942, Geological extrapolation and pseud-abyssal sediments: Am. Assoc. Petroleum Geologists Bull., v. 26, p. 765-792.

Young, L. M., 1968, Boulder beds in Marathon Formation (Lower Ordovician), Marathon Basin, Trans-Pecos Texas (abs.): Geol. Soc. America Spec. Paper 115, p. 381-382.

Plate Section

A. Lower novaculite member on East Bourland Mountain (Loc. 18). Brush line is at contact with Maravillas Formation. Bedding is well defined.

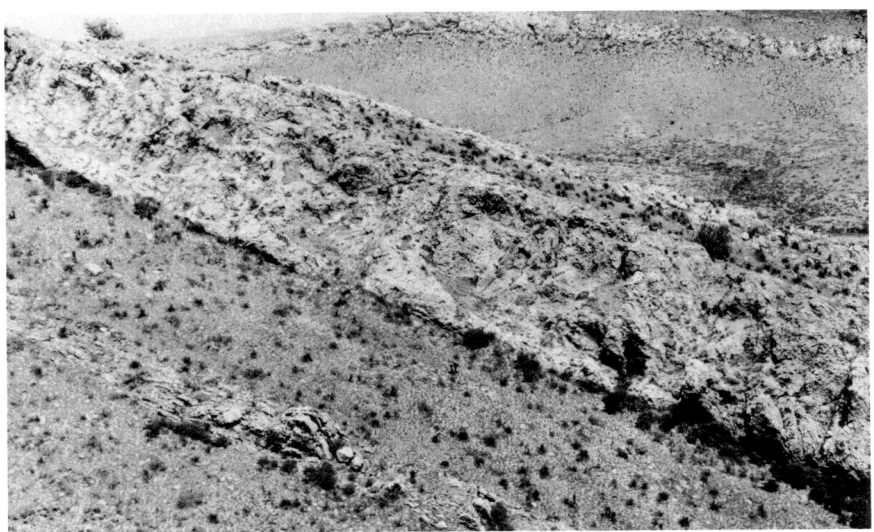

B. Lower three members of the Caballos at Locality 30. Lower novaculite member is the thin resistant unit, lower chert and shale is the grassy interval, the upper novaculite is the thick resistant unit. The lower chert member is absent at this locality.

OUTCROPS OF CABALLOS NOVACULITE MEMBERS

A. Middle two-thirds of the Caballos Novaculite. Thick upper novaculite (left of line) and upper chert and shale members, both with well-defined bedding. Locality 32; view toward the north.

B. Bedding characteristics of lower novaculite member. Locality 14.

OUTCROPS OF CABALLOS NOVACULITE

B. Green chert with partings of siliceous shale in the middle chert and shale member. Conspicuous bedding is typical of this member. Locality 18.

A. Subnovaculite and brown chert with thin shale interbeds. Caballos at the western margin of the Marathon Basin. Locality 6.

CHERT WITH THIN SHALE INTERBEDS

A. Undulose bedding planes in green chert with siliceous shale partings. Close up of area shown in Plate 4B.

B. Stylolites in novaculite bed. Blotchy color is typical of much novaculite. Locality 23.

BEDDING DETAILS OF CHERT AND NOVACULITE

A. Thin bedded brown chert and shale of the Maravillas Formation beneath ledge of thick bedded novaculite of the Caballos Formation. Hammer head is at the contact. Locality 13, picnic grounds.

B. Lumpy upper bedding surface of a slump-deformed chert bed in the upper chert and shale member. Locality 14.

UPPER CONTACT OF THE CABALLOS FORMATION AND SLUMP DEFORMED CHERT BED

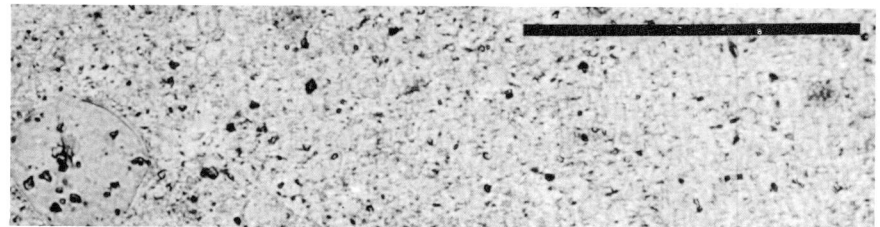

A. (top figure) Novaculite with prominent brown cast viewed in ordinary light. Specks are cloudy dark areas. Minute dark spots include natural hematite and contaminant grinding powder. Scale is 0.2 mm. Sample C-23. Locality 18.

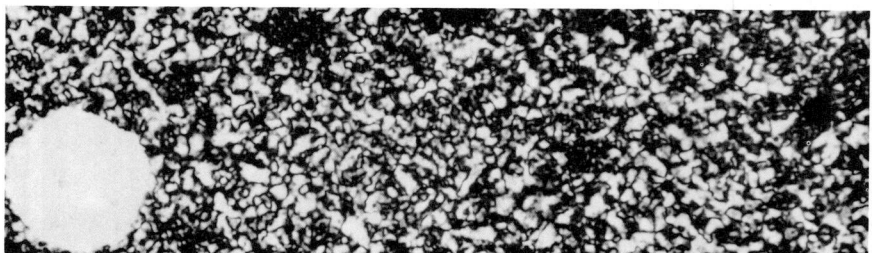

B. Same field as A, crossed polarizers. Clear grain in upper left is detrital quartz. Specks of UFG microquarts occur in MG microquartz matrix.

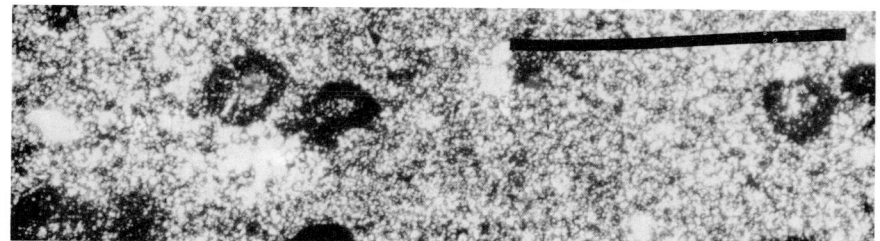

C. Subnovaculite with specks (largely relict Radiolaria) viewed with polarizers crossed. Clear white spots are detrital quartz; matrix of MG microquartz. Scale is 0.5 mm. Sample ST-1, Solitario Uplift.

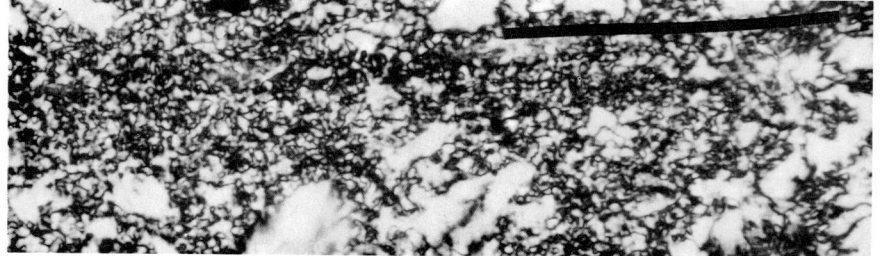

D. Subnovaculite composed of MG-FG microquartz and CG-VCG chalcedony. Polarizers crossed. Scale is 0.2 mm. Sample C-57. Locality 2.

TEXTURE OF NOVACULITE AND SUBNOVACULITE IN THIN SECTION

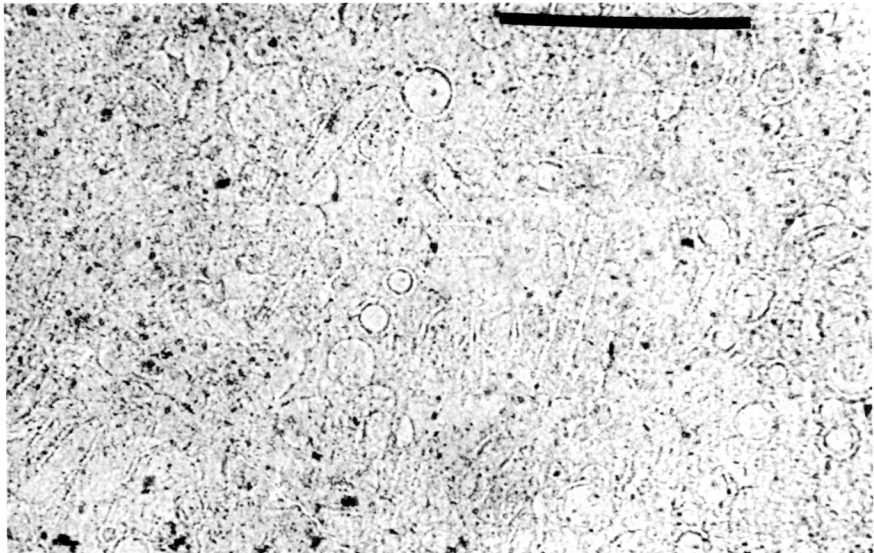

A. Spicule ghosts visible in thin section of novaculite. **Ordinary light.** Scale is 0.2 mm. Sample P-3. Locality 13.

B. Sawed surface of same novaculite sample etched in HF and viewed in reflected light. Most of the recessive areas are relict skeletal grains. Scale same as A.

TEXTURE OF NOVACULITE IN THIN SECTION AND
ETCHED SURFACE

McBRIDE AND THOMSON, PLATE 8
Geological Society of America Special Paper 122

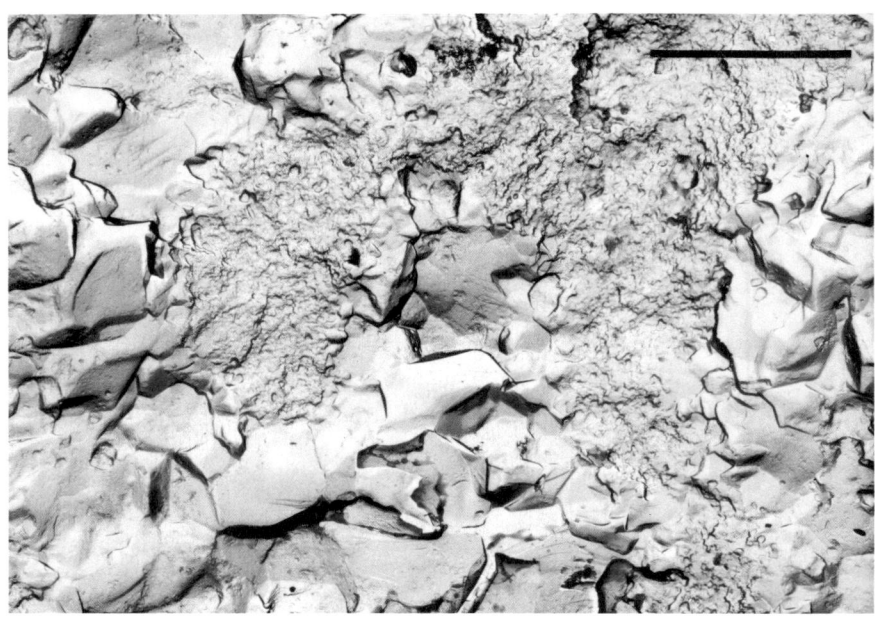

A. Typical bimodal texture of novaculite. See Table 4 for description. Sample C-68. Scale is 10 microns.

B. Hexagonal organic structure of unknown affinity in novaculite. Sample C-P3. Scale is 1 micron.

ELECTRON MICROGRAPHS OF NOVACULITE

A. Fairly even-grained novaculite with circular areas that have radial fabric; the circular areas are probably ghosts of spherical fossils. Sample ZPG. Scale is 10 microns.

B. Coarse-grained mode of sample C-P3 showing polygonal subhedral grains with straight boundaries. Scale is 2 microns.

ELECTRON MICROGRAPHS OF NOVACULITE

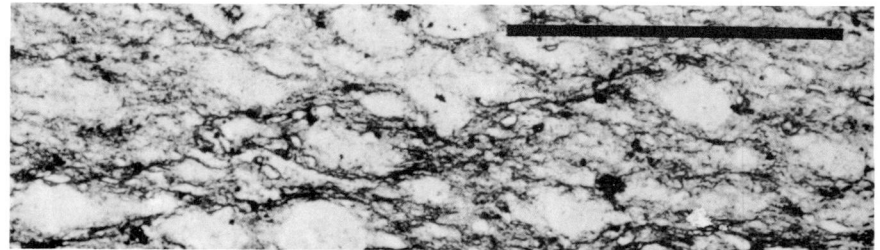

A. (top figure) Green chert with slightly flattened relict Radiolaria. Ordinary light. Scale is 0.5 mm. Sample C-140. Locality 31.

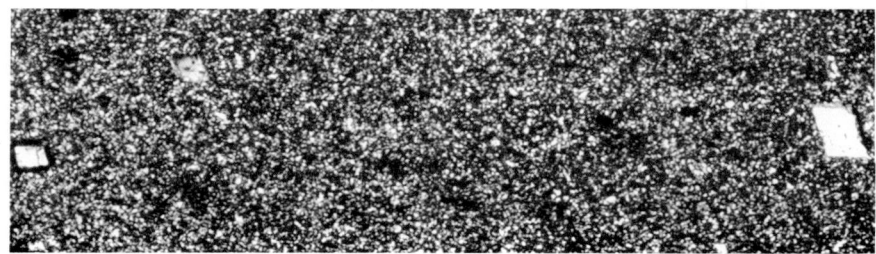

B. Blue-black chert composed of even-grained FG microquartz with scattered rhombs of manganocalcite. Polarizers crossed. Scale same as A. Sample C-81, locality 42.

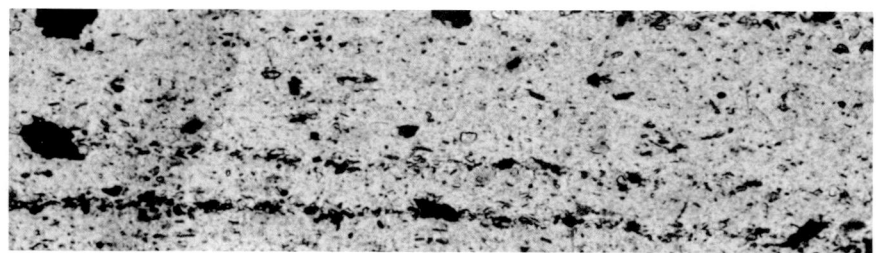

C. Gray chert with silt laminae and scattered grains of oxidized manganocalcite. Ordinary light. Scale same as A. Sample X-63.

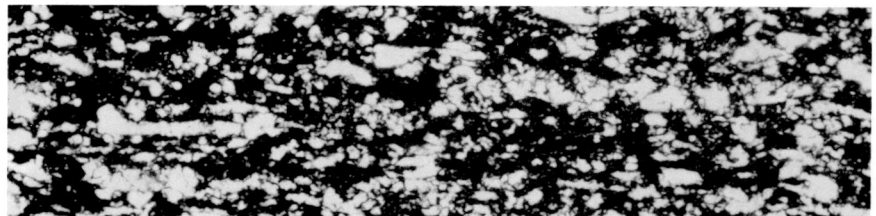

D. Black spiculitic chert. Sample X-50, locality 14. Polarizers crossed. Scale same as A.

TEXTURE OF GREEN, GRAY, AND BLUE-BLACK CHERT IN THIN SECTION

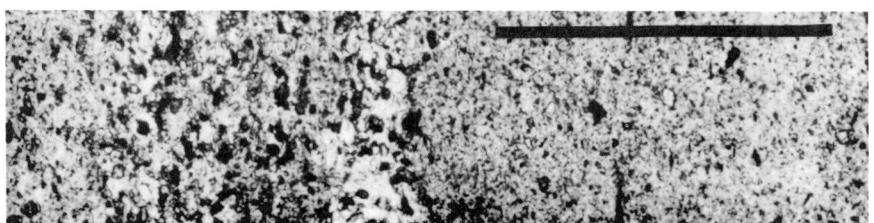

A. (top figure) Gray-green speckled shale with silt laminae. The area by the scale is chiefly illite. Ordinary light. Scale is 0.5 mm. Sample X-59, locality 18.

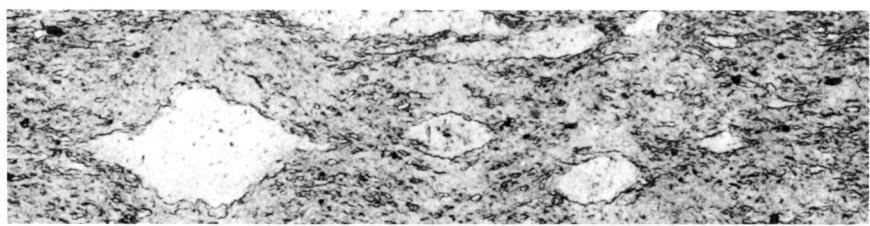

B. Purple-white siliceous shale. Matrix is illite and VFG-FG microquartz that shows differential compaction around flattened Radiolaria. Ordinary light. Scale same as A. Sample X-72, locality 14.

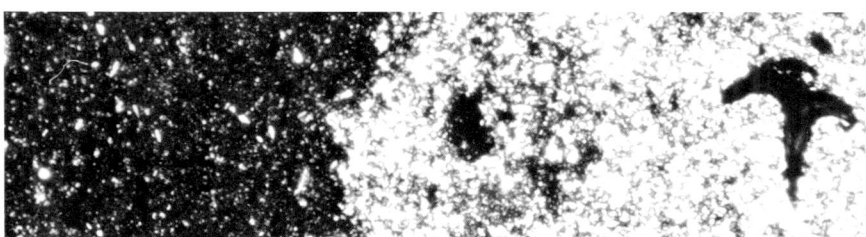

C. Mottled tan chert. Bedding perpendicular to length of photo. Conodont at right. Uneven boundary between UFG microquartz (left) and MG microquartz (right). Polarizers crossed. Sample C-127, locality 28. Scale same as A.

D. Medium-grained quartzarenite with quartz cement. Long black (extinc) grain is phosphatic brachiopod fragment; chert clast at lower right. Polarizers crossed. Sample C-11, locality 14. Scale same as A.

TEXTURE OF SHALE, MOTTLED BROWN CHERT, AND SANDSTONE IN THIN SECTION

A. Horizontal laminations at the top of a graded bed. Scale in inches. Locality 14.

B. Animal trails and burrows on sandstone bedding planes. The upper blocks show upper bedding surfaces, the lower is a sole. Scale in inches. Locality 14.

STRUCTURES IN SANDSTONE BEDS

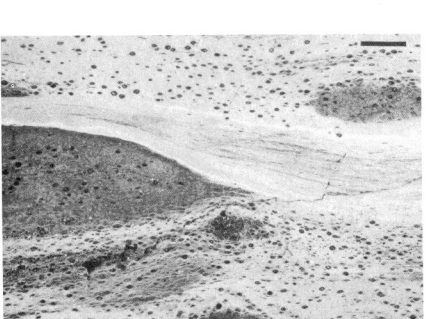

A. Chert pods (dark, unstreaked) in illite (light) and siliceous shale (light, streaked). Radiolaria (specks) show various degrees of flattening. Scale is 1.5 mm. Sample X-74, locality 18.

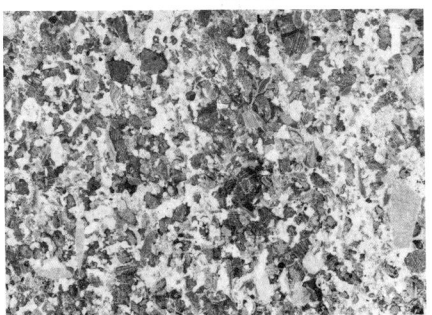

B. Calcarenite with microspar cement. Bedding perpendicular to length of photo. Light grains are micrite rock fragments and intraclasts. Sample C-P8, locality 7.

C. Chertified sandy chert-pebble conglomerate. Bedding perpendicular to length of slide. Sample X-58, locality 14.

D. Novaculite with mottled texture probably produced by burrowing organisms. Polarizers crossed. Sample C-113. locality 20.

Photos by direct enlargement of thin sections, all same scale.

TEXTURE OF RADIOLARIAN CHERT AND SHALE, CALCARENITE, CHERT-PEBBLE CONGLOMERATE, AND NOVACULITE

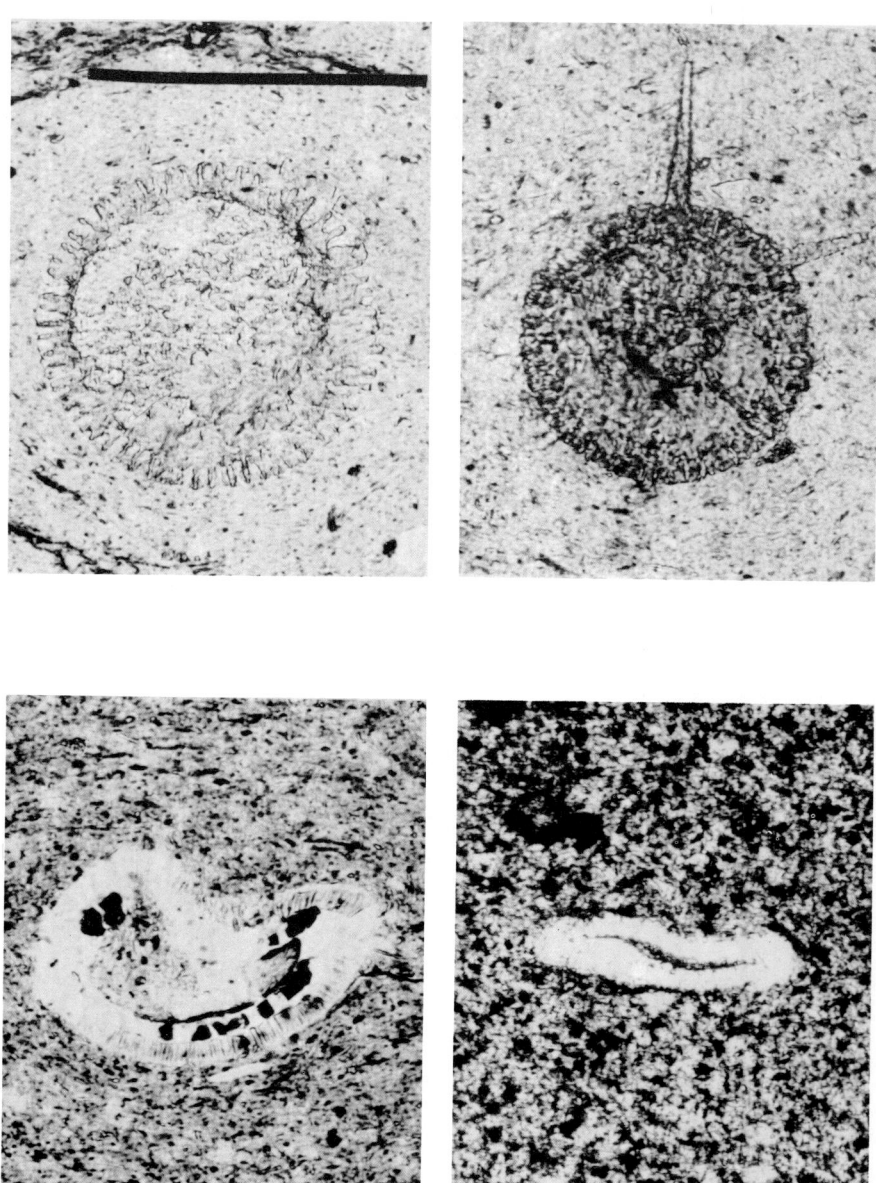

Various degrees of preservation and flattening of spumellinid capsules are visible. All same scale; scale is 0.2 mm. A and B are brown siliceous shale (Sample X-67), C is gray shale (Sample C-34), D is green shale (Sample C-33); all locality 18.

A. Pseudo ripple marks on upper surface of a novaculite bedding plane at locality 23. Amplitude of ripples is 2 cm.

B. High-domed, rounded knobs of diapiric green chert. Top bedding surface. Locality 8.

PSEUDO RIPPLE MARKS IN NOVACULITE AND KNOBS OF DIAGENETIC ORIGIN IN CHERT

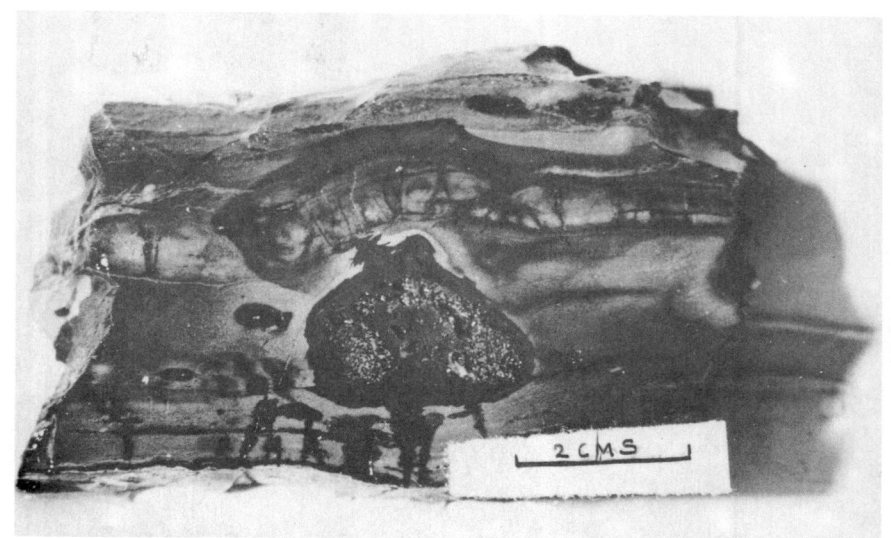

A. Heart-shaped concretion in clayey chert. Dark zones are stained purple and red by hematite. Pores in center of concretion are sites of dissolved crystals, possibly of pyrite. Locality 14.

B. Turban-shaped concretions in lower novaculite member. A central depression is visible on top of the left concretion. Looking down on bedding surface. Locality 18.

IRON-OXIDE CONCRETION AND TURBAN CONCRETIONS

A. Casts of circular mounds, some with concentric rings or central depressions, on sole of bedding plane. Locality 26.

B. Large mounds with central depressions on top of novaculite bed. Locality 17.

PITS AND MOUNDS ON NOVACULITE BEDDING PLANES

Index

Aberdeen, E., 36, 77, 85
Alexander, G. B., 83
Algae, 64
Alpine geosyncline, 78
Ankerite, 64
Arick, M. B., 31
Arkansas
   Novaculite, 10, 31, 36, 40, 41, 50, 52, 64, 65, 75, 76, 78, 92
   State of, 8, 10, 39, 93
Arrhenius, G., 90, 92
Apatite, 62, 80, 82, 83
Aubuoin, J., 34

Bailey, E. H., 75, 89, 91
Baker, C. L., 10, 11, 12, 31, 32, 35, 36, 64, 65, 67, 73, 76, 81, 84, 90
Barghoorn, E. S., 52
Bennett, R. E., 11, 12, 31, 32, 35, 64, 67, 77, 83
Berger, W. H., 85
Berry, W.B.N., 11, 12, 21, 25, 28-30, 33, 35, 40, 94
Bien, G. S., 83
Big Bend National Park, 25
Biotite, 64, 82
Bjorklund, T. K., 24
Blatt, H., 82
Blaylock Sandstone, 40
Bowman, W. F., 10, 12, 31, 32, 35, 36, 64, 65, 67, 73, 76, 81, 84, 90
Brachiopod, 36, 61, 62, 64
Branner, J. C., 77

Breccia, 24, 30, 60, 72, 73, 81, 95
Bryozoa, 64
Burrow, 36, 57, 61, 62, 66, 69, 70, 80
Buser, W., 92

Caballos Novaculite
   age, 31, 36, 37
   composition, 9, 10
   definition, 11, 12
   fossils, 36, 37
   lower contact, 31-34
   members, 11, 12
   origin, 33, 74-92, 94, 95
   stratigraphy, 5-42
   thickness, 9
   upper contact, 34-36
Calcarenite, 18, 20, 37, 64, 79, 80, 86, 93, 94; see also Limestone
Calcite, 64
California, 75
Callixylon, 20, 36, 95
Calvert, S. E., 75
Cambrian, 93
Carver, R. E., 46
Chalcedony 20, 46, 47, 57, 63, 65, 66, 72, 73, 90
Chemical analyses, 52, 58, 62, 88, 89
Chert, see also Novaculite and Subnovaculite
   banded and mottled, 59, 60, 71, 72
   blue and red, 59

brown, 59
clasts, 62-64, 66
definition, 46
description, 52
gray and black, 58, 59
green, 57, 59, 72
Japan, 56
mottled, 20, 58, 71, 72
occurrence, 14-20, 22, 24-30, 33-35, 62
origin, 74, 87, 94
tan, 56, 57
Chlorite, 61, 62
Christie, J., 82
Clarke, F. W., 89
Collophane, 56, 59, 62, 63, 64, 81
Comstock, T. B., 76, 77
Concretions, 66, 68
Conglomerate, 16, 18, 20, 24, 29-36, 63, 64, 66, 80, 93, 94
Conodonts, 35-37, 61, 64
Cotera, A. S., 35, 82
Cretaceous, 9
Crinoid, 64
Croneis, C., 52, 55, 64, 65, 77, 78, 83, 84
Cross section, 21
Cross-stratification, 65
Curray, J. R., 87

Dagger Flat anticlinorium, 19, 20, 22, 24, 40, 59, 66, 67
Daly, J., 5, 39
Davis, E. F., 67, 77, 80, 91
de Laubenfels, M. W., 55, 86
Dennison, R. E., 82
Devonian, 31, 36, 37, 39, 40, 79, 80, 82, 86, 89, 95
Diagenesis, 79-81, 85, 90, 92, 94, 95
Diatom, 79, 83, 91
Dietz, R. S., 87
Dimpled depressions, 20
Dog Canyon, 9, 24
Dolomite, 64, 94
Dove Mountain Quadrangle, 28

East Bourland Mountain, 12, 34, 60, 83
Eifler, G. K., 31
Ellison, S. P., Jr., 35
Eugster, H. P., 75

Fagin, J. J., 71
Fan, P. H., 11, 12, 79
Featherstonhaugh, G. W., 9
Fecal pellets, 36, 51, 77, 83-85
Flawn, P. T., 5, 8, 25, 39, 78, 82, 89, 93
Fluxoturbidite, 34
Folk, R. L., 46, 52, 55, 56, 81, 90
Fort Peña, 16, 32
Fossils, *see* also Algae, Brachiopod, Conodont, Fecal pellets, Gastropod, Graptolite, and Pelecypod)
benthonic, 17
plant, 36
problematic, 25, 28, 36, 37, 55
spore, 36
trace, 17, 36, 66, 69, 70, 87
worm, 87
Fournier, R. O., 83
Franciscan Group, 75, 80, 89, 91, 92
Fusselman Formation, 39, 93

Garrels, R. O., 84
Gas
escaping, 66, 80
pits and mounds, 17, 69, 84, 86
Gastropod, 36
Geopetal fabric, 66, 71
Glass Mountains, 9
Glauconite, 62, 64, 66
Goethite, 57, 73, 92
Goldstein, A., 36, 63, 65, 78, 81, 89
Govett, G. J. S., 75
Graves, R. W., Jr., 31
Griswold, L. S., 9, 75
Grunau, H. R., 74, 83, 88
Grutter, A., 92

INDEX

Gulf of California, 75
Gulf Oil Company well, 39, 93

Haymond Formation, 82
Hazzard, R. T., 25
Heavy minerals, 66, 69
Hematite, 16, 25, 47, 50, 56-68, 61, 72, 73, 81, 85, 88, 92
Henbest, L. G., 36
Hendricks, T. A., 78
Herrin, E. T., 9, 11
Hill, R. T., 10
Holocene, 86, 89, 90, 92
Honess, C. W., 50
Horse Mountain, 11, 24
Hough, J. L., 75
Hudson, F. S., 92

Iler, R. K., 83
Illite, 50, 52, 57, 58, 60, 61, 81, 88
Iron formations, 74, 75
Iron oxide, 62, 71, 73, 94
Isopach map, 21

Jones, T. S., 37

Kaibara, H., 56
Kaolinite, 61
Kennedy, G. C., 83
K-feldspar, 62, 64
King, P. B., 5, 10, 12, 16, 18, 19, 21, 25, 28, 30, 33, 35, 36, 64, 65, 67, 77, 78, 81, 82, 83, 89
Kirchmayer, M., 71
Kirwan, R., 9
Knobby bedding planes, 22, 30, 66, 70, 71
Korbisch, H., 92
Krumbein, W. C., 84
Krynine, P. D., 82

La Berge, G. L., 52
Laminae, 18, 47, 57, 61-65, 80, 85, 91
Liesegang bands, 66, 71

Limestone
  composition, 64
  occurrence, 36, 59
  other comments on, 76, 77, 83, 85
  see also Calcarenite
Limonite, 57, 58, 59, 62, 81
Little Woods Hollow syncline, 71
Llanoria, 82, 93, 95

Magnetite, 88
Manganese oxide, 16, 18, 20, 24, 30, 47, 50, 58-60, 66, 71-73, 81, 92, 95
Manganite, 58
Mangano calcite, 50, 58, 81, 84, 86
Marathon
  anticlinorium, 17, 19, 20, 22, 66, 67
  Basin, 8, 9, 20-24, 30, 32, 39, 63, 66, 70, 82, 93
  geosyncline, 5, 37, 39
  region, 5, 10, 78, 83
  town of, 32, 60
Maravillas
  Chert, 9, 16, 25, 28, 29, 31-34, 64, 83, 93, 94
  Gap, 24
Martite, 73
Maxwell, R. A., 25
McBride, E. F., 31, 35, 37, 45, 82, 87, 93, 94
McGlasson, E. H., 39, 93
Measured sections, 5, 20, 21
Megaquartz, 46, 47, 56, 60, 66, 68, 71
Mesozoic, 80, 86
Microquartz, 46, 47, 50, 52, 55-58, 60, 61, 63-65, 71, 79, 81, 84, 90-92
Midland Basin, 39
Miser, H. D., 75, 92
Mississippian, 35, 37, 40, 80, 90, 95
Missouri Mountain Shale, 40
Mixed-layered clay, 62
Mizutani, S., 91
Molluscs, 64
Montmorillonite, 61

Moore, D. G., 87
Mounds, 66, 69

Nannoplankton, 55
Nevada, 71
New Mexico, 39, 93
Nielsen, H. M., 11, 12, 21, 25, 28-30, 35, 40
Novaculite
  definition, 10, 46
  description, 46-56
  occurrence, 16-19, 22, 24, 25, 28-30, 40, 57
  origin of rock, 74-92, 94, 95
  origin of term, 9, 10

Oklahoma, 8, 39, 93
Old Jones Ranch, 9, 28, 32
Ordovician, 9, 31, 33, 37, 93, 94
Organic matter, 57-59, 61, 65, 81, 84, 87
Ostracods, 64
Ouachita
  geosyncline, 5, 14, 39, 40, 75, 78, 93
  Mountains, 8, 40, 75

Paleozoic, 5, 9, 14, 29, 56, 78, 82, 93
Palinspastic map, 21
Park, D. E., Jr., 52, 55, 64, 65, 75, 78, 83, 84
Payne Hills, 62, 66
Pelecypod, 36
Pelmatozoan, 64
Pennsylvanian, 82, 93, 95
Persimmon
  Gap, 9, 24, 29, 32
  Gap Formation, 25, 29, 32
Petrified wood, 20, 36, 59, 83
Pettijohn, F. J., 74
Pillow structure, 66
Pits, 66, 69
Plant remains, 36
Powers, S., 11
Precambrian, 51, 74, 75
Pseudo ripple marks, 66, 67

Purdue, A. H., 75
Pyrite, 50, 56, 57, 58, 73, 81, 92
Pyrolusite, 58, 92

Quartz, 50, 56, 60-65, 77-79, 82, 83, 90, 93
Quaternary alluvium, 9

Radiolaria, 36, 37, 47, 50, 51, 55-58, 61, 65, 66, 74, 77, 78, 80, 83-85, 90-92, 94, 95
Rhodochrosite, 50
Ripple marks, 11, 17, 66, 73, 77
  see also pseudo ripple marks
Rock
  fragments, 62
  House Gap, 63
Ross, C. S., 39, 93
Rough Creek, 9, 24, 25, 28, 29, 32
Rutile, 62

Sandstone
  description, 62, 63
  occurrence, 17, 18, 24, 37, 93
  origin, 80
Santiago
  Chert, 11, 30, 36
  Mountains, 9
Schoolcraft, H. R., 9
Schoonover Formation, 71
Seilacher, A., 87
Sellards, E. H., 11, 31
Shale
  clasts, 62, 63
  description, 60-62, 70
  occurrence, 14-20, 22, 24, 25, 30, 32, 59
  origin, 80, 88, 89, 94, 95
Siderite, 63
Siever, R., 74, 83
Siliceous
  ooze, 74, 75, 77, 81, 84, 86, 87, 90, 91, 94, 95
  shale, 14, 16, 17, 18, 20, 22, 28, 29, 32, 34, 57, 60, 73, 79, 95

Index

Silicoflagellate, 80, 91
Siltstone, 18, 62, 64, 80
Silurian, 31, 37, 39, 41, 80, 82, 86, 89, 94
Slick-Urshel Oil Company well, 39, 93
Sloss, L. L., 41
Soft-sediment deformation, 20, 34, 59, 60, 65, 66, 72, 95
Solitario
   region, 5
   Uplift, 9, 11, 28, 30, 32, 34, 39, 66, 93
Specks, 47, 50, 51, 65, 66, 92
Sponge spicules, 36, 37, 47, 50, 51, 55-59, 61, 65, 80, 85, 86, 90, 91, 94
Spore, 36, 65
Sterling, P. J., 41
Stratigraphy
   lower chert member, 12, 14, 16, 22, 33
   lower chert and shale member, 12, 17, 19, 22
   lower novaculite member, 12, 16, 17, 22
   upper chert and shale member, 14, 19, 20, 22
   upper novaculite member, 12, 19, 22
Stride, A. H., 90
Stylolites, 14, 19, 30, 47, 67-69, 73
Swallow, J. C., 90
Subnovaculite
   definition, 10, 46
   description, 56
   occurrence, 14, 16-19, 22, 24, 30, 33
   origin, 74-92, 94, 95

Tabosa Basin, 39, 93
Taliaferro, N. L., 91, 92
Tertiary, 83
Tethys geosyncline, 74, 88
Tesnus Formation, 9, 25, 29, 35-37, 93, 95
Thomson, A., 28, 31, 32, 35-37, 45, 65, 82, 87, 89, 94
Tourmaline, 62
Trans-Pecos Texas, 5, 14, 39, 93
Trilobite, 64
Turbidite, 40, 80-82, 86, 94
Tyler, S. A., 52

Udden, J. A., 10, 31
Ulrich, E. O., 31
Unconformity, 31, 33, 35

van Lier, J. A., 83
van Waterschoot van der Gracht, W. A. J. M., 11
Volcanic
   ash, 76, 78, 80
   rock, 64
von Streeruwitz, W. H., 10

Weaver, C. E., 46, 52, 55, 56, 81, 90
Well data, 39
Wheeler, H. E., 41
Wilson, J. L., 11, 25, 28, 29, 32, 39, 93
Woodford Formation, 39
Woolnough, W. G., 86
Worthington, L. V., 90

Young, L. M., 93

Zircon, 62, 80, 83